T0269640

# PLANNING A CAREER IN BIOMEDICAL AND LIFE SCIENCES

# PLANNING A CAREER IN BIOMEDICAL AND LIFE SCIENCES

## MAKING INFORMED CHOICES

**AVRUM I. GOTLIEB**
Professor, Laboratory Medicine and Pathobiology
Senior Academic Advisor to the Dean of Medicine
Faculty of Medicine, University of Toronto
Laboratory Medicine Program, University Health Network
Toronto, Ontario, Canada

AMSTERDAM • BOSTON • HEIDELBERG • LONDON
NEW YORK • OXFORD • PARIS • SAN DIEGO
SAN FRANCISCO • SINGAPORE • SYDNEY • TOKYO
Academic Press is an imprint of Elsevier

Academic Press is an imprint of Elsevier
32 Jamestown Road, London NW1 7BY, UK
525 B Street, Suite 1800, San Diego, CA 92101-4495, USA
225 Wyman Street, Waltham, MA 02451, USA
The Boulevard, Langford Lane, Kidlington, Oxford OX5 1GB, UK

**Notices**
Knowledge and best practice in this field are constantly changing. As new research and
experience broaden our understanding, changes in research methods, professional practices,
or medical treatment may become necessary.

Practitioners and researchers must always rely on their own experience and knowledge
in evaluating and using any information, methods, compounds, or experiments described
herein. In using such information or methods they should be mindful of their own safety
and the safety of others, including parties for whom they have a professional responsibility.

To the fullest extent of the law, neither the Publisher nor the authors, contributors, or
editors, assume any liability for any injury and/or damage to persons or property as a
matter of products liability, negligence or otherwise, or from any use or operation of any
methods, products, instructions, or ideas contained in the material herein.

ISBN: 978-0-12-802242-9

**British Library Cataloguing-in-Publication Data**
A catalogue record for this book is available from the British Library

**Library of Congress Cataloging-in-Publication Data**
A catalog record for this book is available from the Library of Congress

For Information on all Academic Press publications
visit our website at http://store.elsevier.com/

Typeset by MPS Limited, Chennai, India
www.adi-mps.com

Printed and bound in the United States of America

Working together
to grow libraries in
developing countries

www.elsevier.com • www.bookaid.org

# DEDICATION

Dedicated to my wife Linda and to the memory of my parents, Harry and Roberta Gotlieb.

# CONTENTS

# MY CAREER IN BIOMEDICAL AND LIFE SCIENCES

I obtained my BSc in Psychology and Physiology with first class honors (1967) and my MDCM (1971) from McGill University. I continued my training in medicine and anatomic pathology at the teaching hospitals of McGill University. I obtained my fellowship from the Royal College of Physicians and Surgeons of Canada in Anatomic Pathology (1975) and certification from the American Board of Pathology (1976). I pursued research training in cell biology in the  Department of Biology, University of California San Diego with Professor S.J. Singer, supported by a Medical Research Council Fellowship. I then ran a vascular cell biology research lab at the University of Toronto for 35 years. During that time, I was the founding Chair of the Department of Laboratory Medicine and Pathobiology (LMP), Faculty of Medicine, University of Toronto (1997–2008). I was appointed Interim Vice-Dean, Research and International Relations (2009–10) and Acting and Interim Vice-Dean, Graduate and Life Sciences Education (2011–2014) in the Faculty of Medicine. Currently I am Senior Academic Advisor to the Dean, Faculty of Medicine, University of Toronto (2010–2015).

My research interests include atherosclerosis and valvular heart disease. I have published on blood vessel repair, especially on the role of the cytoskeleton in endothelial repair, and I also studied how heart valve cells repair valves after they have been injured. I published over 100 peer reviewed papers, and 35 reviews and book chapters. I edited three books, including the comprehensive textbook *Cardiovascular Pathology*, edited with colleagues M.D. Silver, University of Toronto, and F. Schoen, Harvard Medical School. I have received peer reviewed funding from the Heart and Stroke Foundation of Ontario and the Medical Research Council, now

Canadian Institutes of Health Research (CIHR). Many of my graduate students also received studentships from outside the university to carry out their graduate training. After leaving my laboratory they followed diverse career paths in academia, the pharmaceutical and biotechnology industry, medical practice, physiotherapy and anatomic pathology, among others.

I am the past coeditor of *Cardiovascular Pathology*, a journal of the Society for Cardiovascular Pathology dedicated to basic, clinical, and applied cardiovascular science published by Elsevier. I serve on the editorial board of *The American Journal of Pathology* (AJP) and of Laboratory Investigation.

I am a former President of the American Society for Investigative Pathology (ASIP) and past President of the Canadian Society of Atherosclerosis, Thrombosis and Vascular Biology (CSATVB) and the Society for Cardiovascular Pathology (SCVP). I was a member of the board of the Federation of American Societies of Experimental Biology (FASEB) and served as FASEB Vice President for Science Policy. I am an elected Fellow of the Canadian Academy of Health Sciences and a Senior Fellow of the Association of Pathobiology Chairs (APC). I was honored by SCVP with the Distinguished Achievement Award and by APC with their Distinguished Service Award.

This book is a natural outcome of my lifelong interest in ensuring that students make informed choices so that they gain the most out of every stage of their training and find the most suitable and fulfilling careers. In the early stages of my career, this impulse found expression in my directing undergraduate and graduate postsecondary education programs. I was Coordinator of Graduate Studies in Pathology, responsible for training primarily nonclinical students, and I was Course Director of the Pathobiology course in Undergraduate Medicine, which was part of the medical school curriculum. In 2000, along with several colleagues, I initiated an innovative and unique undergraduate arts and science Specialist Program in Pathobiology. The goal was to introduce non-MDs very early on in their life science training to the research world of mechanisms of disease and help them make informed career choices. The program offered an extensive number of introductory and advanced courses in the biology of human diseases as well as hands-on research laboratory opportunities. ASIP honored me with the Robbins Distinguished Educator Award for these activities in 2011, and I have presented career talks at several national and international scientific meetings and venues.

# PROLOGUE

The book begins by addressing the high school student and guidance counselor and continues with career guidance spanning undergraduate, graduate, and postdoctoral studies. The book focuses on how best to plan for a successful career in biomedical and life sciences, beginning with high school and ending with job hunting, establishing one's self in a position and achieving tenure and promotion.

The objective of this book is to provide a competitive advantage to students and junior faculty by presenting useful information, insights, and tips to guide them on their journey of career choice and career development in the biomedical and associated life science professions. The book begins with fostering interest in high school and presents career guidance through undergraduate, graduate, and postdoctoral studies. The book focuses on how best to plan for a successful career in biomedical and life sciences, including job searching, establishing one's self in a position and achieving tenure and promotion. These plans, if adhered to, provide the student with the opportunity of making informed choices that will allow their training, job search, and professional experiences to guide them to the most suitable and fulfilling careers in both the academic and nonacademic worlds, including industry, business and life sciences related professions. This focus on both sectors is important since both are competitive job markets that attract life sciences graduates. Much has been written on career development; however, with the dramatic increase in career choices available for those training in the biomedical and life sciences, students and faculty need to begin planning early to meet the challenges of a dynamic training path and job market. Although trained in research, the graduates will begin careers in many different areas. Some will remain in academic or industry research, others in science nonresearch jobs and others in non-science, nonresearch jobs. It is the research training however that propels them into the workforce.

When I first began my undergraduate studies and throughout my training, I was fortunate to be blessed with caring mentors who helped me along the way. Professors Shao-nan Wong, Robert Moore, David Kahn, Jon Singer, Emanuel Farber, Malcolm Silver and Arnie Aberman, among others, are due my eternal gratitude. Their input and concern helped me discern, plan, and achieve my goals. In this book, I would like to partially repay my debt to them by providing students, junior faculty, and/or employees with my hard-won knowledge and experience based on over 35 years as an academic leader in life sciences and medicine at one of the world's great universities, the University of Toronto, and at its outstanding network of teaching hospitals and associated research institutes.

As Chair of Laboratory Medicine and Pathobiology and in other administrative positions, I have already had the good fortune to guide students into successful careers in both the academic and nonacademic worlds of industry, business and life science related professions. However, this book is not only useful for those climbing the career ladder but also for faculty mentors and supervisors, called upon to give advice in an environment so different than the one they were trained in. In addition, this book will be very useful to high school science teachers and guidance counselors to guide students on the principles and strategies of career development in biomedical and life sciences research and teaching.

I would like to acknowledge the valuable discussions and interviews with numerous colleagues, trainees, and students who shared their thoughts with me on this important topic over the past 35 years. Their insights into academic training, scholarship, and mentorship are much appreciated. I appreciate the helpful comments and discussions on nonacademic careers in life sciences with Songyi Xu, a graduate student in Laboratory Medicine and Pathobiology and Co-President of the Life Sciences Career Development Society, University of Toronto. And finally, I thank my own mentors who were willing to devote time and energy to help me make informed choices and shape my training and career, as they did for so many others.

I appreciate the insights on graduate and postgraduate education of Professors Michelle Bendeck and Jeffrey Lee, Department of Laboratory Medicine and Pathobiology, University of Toronto, and the administrative assistance of Bessie Gorospe, Graduate and Life Sciences Education, and Heather Seto, Laboratory Medicine and Pathobiology, University of Toronto. I am grateful to Meshulam Gotlieb for his professionalism and imagination in helping me develop and further clarify important aspects of the book. His wonderful ideas and helpful comments contributed significantly to the overall tone and style of the text.

**Avrum I. Gotlieb**
December 2014

## Fundamental Principles and Components of a Successful Career in Life and Biomedical Sciences

- Be well informed.
- Plan ahead but seize opportunities.
- Choose Mentors.
- Choose Supervisors and Bosses well.
- Work hard and work smart.
- Budget your time carefully and set priorities.
- Be well trained, innovative, and learn to assess risk.
- Communicate well.
- Respect research integrity.
- Stay focused with a strategic plan but be open to new directions.
- Be collegial and be a team player.
- Network and be active in your scientific community.
- Be a lifelong learner.
- Pay attention to and nurture supportive relationships, family, and friends.
- Acknowledge and thank those who help you on the way to success.
- Have fun and keep your sense of humor about you.

# CHAPTER 1

# Getting Bitten by the Bug

## Contents

## Summary

It is not uncommon for students to develop an early interest in the life sciences and it is never too early for them to consider a career in this field, even as early as high school. In fact, prospective life science students should begin seeking information and asking questions early in order to be well informed to make decisions on careers and to avoid regrets later on. They may begin by speaking with high school guidance counselors, searching for information on the Internet, and visiting university information fairs. Students should ensure that their information on entrance requirements, programs, and scholarships is up to date since this information does change frequently. It is also important for students to consider any barriers that may prevent them from obtaining postsecondary education and to understand how they may overcome these barriers. Pipeline programs should be sought out by students to help gain insight into university life and admissions as well as make contacts with mentors who can guide them in making informed career choices.

## 1.1 HIGH SCHOOL—A PLACE TO START

If you have a keen interest in how living organisms work, enjoy the pursuit of new knowledge, and are excited by new discoveries, biological sciences may be the career for you. High school is not too young an age to begin considering a career in life sciences. Many of my colleagues developed an early passion for biology because of their desire to understand their surroundings and were fortunate to be encouraged by teachers who opened their minds to the questions posed by biology and to the pleasure of solving scientific riddles. Like them, you may have been intrigued by biology classes and assignments on how the cells and the molecules inside and outside the cell control human biology. You may have surfed the Internet for life science sites that explain how tissues, cells, and molecules organize the many functions of microorganisms, insects, animals, and humans. You may have seen videos of live cells moving and growing and you may have manipulated three-dimensional structures of molecules online. As you have come to realize that life and biomedical

*Planning a Career in Biomedical and Life Sciences.*
DOI: http://dx.doi.org/10.1016/B978-0-12-802242-9.00001-3

sciences provide you with the keys to unlock and discover the workings of the human body, you have become more and more excited by this field.

Now in senior high school you want to venture into the unknown of life science and biomedical research. You want to understand the molecular mysteries of the human organism, so that you can delve into the mysteries of human existence or use science to prevent or cure disease and human suffering. University students or professors may have visited your high school to explain interesting concepts in biology and share their excitement about new discoveries that change how we think about biology, health, and disease. Perhaps, your science teacher has coached you in a science fair. Perhaps, you have already had the opportunity to do hands-on research in a professional laboratory or had mentored opportunities to carry out discovery research.

Now it is time to take the next step: speak to your school's career guidance counselor, scan the Internet, and visit research exhibits and university information fairs to obtain information about careers. On my own campus, information days for local high school students are well attended. Our program directors and students field questions all day long and information brochures rapidly disappear from our information booths.

Note that your high school should present such studies as a realistic option for you and your friends, especially targeting underrepresented and underserved groups. If you come from a family or cultural background that has not traditionally encouraged attendance at college/university, or if you and your family do not have the financial means necessary to pursue postsecondary education, this encouragement is especially important. Educational pipelines that provide encouragement and support for under-privileged and diversified high school students are available, and should be accessed by you if they are helpful to you for career development.

In many instances, those who need assistance preparing for obtaining a postsecondary education can receive help from targeted local, provincial, state, and federal programs. You must explore these fully with the help of career guidance counselors. These programs are usually provided by individual schools and boards of education and by government agencies. Many colleges and universities also offer programs directed at less advantaged high school students to provide them with an opportunity to meet with active faculty and to attend and participate in lectures, seminars, and laboratory experiences. This offers them a glimpse of what the future may hold, provides them with opportunities to establish contacts with mentors, and allows them to be better informed and plan ahead.

## 1.2 EARLY GUIDANCE

The field of life sciences is broad and making choices on how to enter the area may be initially bewildering. It is up to you to actively seek out information. Do not be shy. Be polite but be persistent so that you obtain the answers you need. Knowledgeable career counseling, especially from high school and university counselors and undergraduate and graduate program leaders, can help you discover what is available and what prerequisites and eligibility criteria exist.

> **Make Informed Choices**

Guidance starting as early as high school should be sought. Guidance is just that. It does not commit you to a particular course of action; however, it provides you with valuable information and allows you to make informed choices. Because entrance requirements, programs, and scholarship opportunities do change, make sure you keep your information current.

The path you are choosing today begins in high school, continues in university on an undergraduate level, and leads into graduate and/or professional schools. After completing these three phases, further postgraduate and postprofessional training (may) await you. You may enter or leave the path at any stage, and career changes or changes in interests are not uncommon at all. To remain motivated throughout this long process, you must have a passion for what you are studying, have fun doing the work, and realize the value in your research, both on a personal level and for society.

## 1.3 AVOID REGRETS

Comments I frequently hear from both students and colleagues are "if only I would have known," "I missed the deadline to apply," and "if only someone would have told me to do this at the time, my career path would have been much smoother." These sentiments can be avoided if you spend the time and energy to gather information on career development. Furthermore, you will have fewer regrets if you sharpen those skills that apply to any career path, such as resume writing and interview techniques, and you acquire specific information about your own areas of interest, such as the best schools to train at and the best journals to read. Some of the information you need is obvious, but to your surprise, you will discover that by being attentive, crucial information about seemingly unimportant topics will also come your way. You cannot find the answer to a question that you did not even pose in the first place. Thus, to be successful, start very early to immerse yourself in career development so that you are able to think and plan ahead.

## CHAPTER NOTES

_____

_____

_____

_____

_____

_____

_____

_____

_____

_____

_____

_____

_____

_____

_____

_____

_____

_____

_____

_____

_____

_____

_____

_____

_____

_____

_____

_____

_____

_____

_____

_____

_____

_____

_____

_____

_____

_____

_____

_____

_____

_____

# CHAPTER 2

# Creating a Dynamic Training Path

## Contents

## Summary

After researching information about careers in life sciences, students must now develop a written plan for achieving their career goals. This typically involves a combination of short- and long-range planning. It is important to understand that plans may change as interests and opportunities change; therefore plans will need to be revised and modified often. Furthermore, it is essential that the plan be flexible in case of disruptions and to take advantage of serendipitous opportunities. Remaining well informed of options will assist students greatly with revising their plan. The lifestyle of an individual in the life sciences is highly desirable as it allows the individual to be immersed in a fast-paced and ever changing environment. It is hard work but it is also exciting as you join the global community of like-minded scientists who expand the knowledge and applications of biomedical and life science to research, business, and professional careers.

## 2.1 PATHWAYS TO SUCCESS

A training path refers to the steps you choose to take out of a variety of possibilities to achieve the training objectives you set for yourself. Usually it is a combination of short- and long-range planning. You should focus on acquiring very high-quality training in a timely fashion. Some students thrive in programs that offer considerable flexibility to achieve the best training while others do well in more structured programs. Ultimately, your training will be the result of a carefully crafted long-range career plan, often referred to as an individual development plan (see Chapters 3–5 and 7), and serendipitous opportunities that appear as you follow your training path. These opportunities seem to appear by chance, but as Louis Pasteur said "Serendipity favors the prepared mind." Surprised though prepared for such an outstanding opportunity, you will have to decide whether to take this detour from your path or not. By treating your plans as dynamic and subject to change, though not on a whim, you can evaluate such opportunities in collaboration with your advisors and mentors. In any case, you should revisit your plan every six months.

Within the rubric there are several different pathways that you may take to reach your ultimate goal: some traditional and some novel

*Planning a Career in Biomedical and Life Sciences.*
DOI: http://dx.doi.org/10.1016/B978-0-12-802242-9.00002-5

involving interludes and diversions that help you define your goals by being exposed to a variety of experiences in and out of science. The traditional path is high school biology followed by undergraduate studies with an emphasis in life sciences that include a hands-on laboratory research experience that may take the form of research course projects. Summer research in a laboratory is highly regarded. Following graduation, graduate school and postdoctoral training are what

| Planning and Serendipity |
|---|

comes next in sequence. Having taken this solid path, you are well prepared to take on an academic or a nonacademic job. However, you will find that even in taking this traditional path, your own personal circumstances are important in determining the steps you take along the way. Remember that although the outcome is important, the journey must be pleasant and enjoyable, but as direct as possible, so that you do not become a perpetual trainee. Furthermore, while trying to establish a deliberate career pathway based on the best information available, remember that flexibility is necessary to respond to chance opportunities that appear from time to time.

Career pathways may also be more novel including breaks along the way. This usually includes stepping out of university and working for a while. Such a break will provide income to support future studies, and it will provide the change of environment necessary to allow you to gain a better perspective on where you are heading. Such a time-out is not uncommon. Sometimes the time-out still keeps you in a research laboratory environment as a technician, but sometimes it allows you to explore other spheres in science or outside of science. You may travel the world but attempt to maintain a career focus in part by choosing to use some of your time-out to travel to other locations and explore future study opportunities.

One of our students was set on a well-trodden path to enter a specific discipline. She had been moving in this direction for several years. She took the opportunity between semesters to carry out an international volunteer placement in a third-world country which opened her eyes to new career opportunities that she did not even know existed. She then took a year off to pursue her new work. Upon her return, she discussed her experience with her mentors, carefully gathered as much information as possible, and after weighing the pros and cons, altered her course of study to provide her with the required knowledge to pursue her newly found passion.

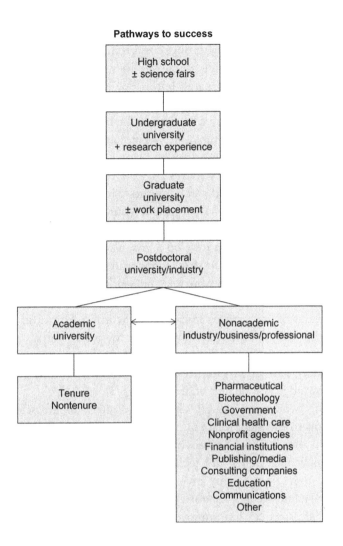

Pathways to success

## 2.2 LIFESTYLE AND COMMUNITY STATUS

The lifestyle is desirable. You work in an extremely fast-paced and competitive environment whether you are in the academic or nonacademic spheres. Life sciences tend to attract people of great intelligence and creativity. In addition to science, many are also immersed in life-enriching pursuits in music, art, and literature. This is a very exciting environment to be in. Stimulating professional conversations often revolve around research challenges to be overcome and novel experiments to try. These discussions are full of energy as new ideas to tackle issues flow freely.

I have found a very collegial community in life science and biomedical research. In the laboratory group, conversations always spill over into local and world issues. We avidly discuss politics, religion, health care, poverty and the income gap, to name but a few. The arts, the latest play in town, the newest film, and, of course, sports are also firmly on our radar. Although I live in a hockey city, baseball, basketball, and football also catch our interest. At World Cup time, the debate about soccer was especially intense. This stimulating environment and true collegiality are but some of the benefits of a career in life sciences.

Although the work is hard and consumes a lot of time and energy, you have the opportunity to travel to present your findings to colleagues at universities, research institutes, and scientific meetings around the globe. Through this interaction, you develop a group of like-minded friends and acquaintances with whom you share wonderful experiences. For many of my colleagues, the opportunity to see the world and interact with some of the top minds in the international scientific community make the trials and tribulations of life science's research well worth it. For most, the career journey in life sciences is fun and immensely rewarding, for as a life scientist, you are at the forefront of biology and of the knowledge and innovations economy, part of an international like-minded fraternity advancing human knowledge and quality of life.

As if this were not enough, choosing a career in life sciences also commits you to life-long learning even once you have settled into a job. Everyone in the discipline is constantly learning new ideas and concepts as well as new techniques and technologies. Being a life-long learner is an essential ingredient for my own success and also provides me with a contemporary knowledge base for civic engagement. I have found that my training in life sciences has improved my capacity to carry out my own professional work as a pathologist and laboratory physician. Your knowledge similarly can and should propel you outside the life sciences sphere and into the community. There, you can provide guidance and leadership to many civic causes that will benefit greatly from your unique knowledge.

For many in the discipline, the research is ultimately focused on solving questions in human life sciences, although the domain of life sciences is very broad and includes all types of biology, botany and microbiology, to name but a few. Such a specialization gives you insight into many biological and biomedical issues that you come across in daily life. On a personal note, you will understand how your own body functions, in sickness and in health, and how environmental agents may cause injury or disease.

You will have a better understanding of how particular lifestyles can benefit or harm your body. You will understand nutrition better than your neighbors. Ultimately, a career in biomedical and life sciences is more than a job; it is a lifestyle, an exciting and challenging lifestyle that is satisfying and rewarding at the professional and personal levels. As you immerse yourself in this lifestyle, you should actively join the international community of scientists through collaborative research, topic specific networking, and active participation in national and international scientific and discipline societies that foster research, training, career development, and education.

## CHAPTER NOTES

# CHAPTER 3

# Undergraduate Studies

## Contents

## Summary

When beginning your undergraduate education you will be faced with the dilemma of depth versus breadth training. A combination of the two is desirable as students will then possess both the in-depth knowledge required and the ability to collaborate with colleagues from other disciplines and look at biomedical problems through different lenses. Specialist programs, focused on in-depth research and education, provide excellent platforms for career development. During undergraduate studies, it is critical that you gain hands-on undergraduate research experience in order to be accepted into a graduate life science program. Mentored undergraduate research opportunities should be embraced. In addition to helping your graduate school application, engaging in research will also provide you with the opportunity to develop laboratory, communication, and life-long learner skills and to network with peers and scientists from a variety of institutions. Aside from family, friends, and partners, it is strongly advised to seek out mentors for guidance and support throughout your training pathway to success. Identify suitable mentors for yourself to provide career guidance as you proceed with your training. Mentorship should be a two-way street in which the mentee and the mentor gain from this unique faculty–student relationship. Mentors are important in both academic and nonacademic environments.

## 3.1 DEPTH VERSUS BREADTH IN YOUR UNDERGRADUATE CURRICULUM

You are going to be faced with options for breadth and depth training as an undergraduate student. The former provides more introductory and survey courses and is meant to expose you to several areas of study to provide you with a rounded education. The latter provides sufficient study in a focused area to provide you with content expertise and usually provides you with the opportunity to do research projects and/or in-depth analysis of a topic resulting in a mini thesis of sorts.

> Specialist Programs Provide Excellent Platforms for Career Development

*Planning a Career in Biomedical and Life Sciences.*
DOI: http://dx.doi.org/10.1016/B978-0-12-802242-9.00003-7

On the one hand, the breadth route has the advantage of making the student a better candidate in the future for carrying out research at the interface of disciplines. Since interdisciplinary research is encouraged in biomedical and life sciences to investigate and solve complex problems, especially those of a global nature, such training provides a certain advantage. Indeed, students are encouraged to learn to work across disciplines and to be exposed to transdisciplinary and team approaches to problem solving, so breadth has its place. On the other hand, the in-depth route has proven itself to be an excellent testing ground for potential success in graduate studies since it challenges the student academically and shows potential supervisors that the student has the capacity to explore problems and seek solutions. The in-depth approach also has the advantage of endowing students with skills that are transferrable to other areas of study or different career pathways should they choose a different focus or path.

Some students elect to strive for a combination of the two routes by taking, for instance, a major program in one area to provide breadth and a specialist program in another to provide depth. The combination of breadth and depth is obviously optimal since students are later expected to bring their in-depth knowledge to the table to tackle problems in tandem with their colleagues from other disciplines. In this process, the problem is examined from many perspectives, but each discipline makes its own unique contribution.

The relatively recent innovation of linking undergraduate and graduate student education provides new transformative opportunities for advancing student education. Be sure to take advantage of this. Likewise, linking education and research has resulted in course offerings with more depth and experiential research opportunities. Today, undergraduate students have an easier time gaining exposure to the innovation agenda of graduate education. This results in the best research faculty being available to teach and mentor undergraduates, and undergraduate students have greater interaction with graduate students, thus facilitating useful mentorship relationships. Nationally and internationally, departments engaged in cutting-edge research develop online learning opportunities that open up new doors for undergraduates. Furthermore, linking of advanced senior undergraduate learning with graduate education provides undergraduates with the in-depth scholarship necessary for many career paths and for embarking on further advanced education.

## 3.2 MENTORSHIP, AN ESSENTIAL INGREDIENT FOR SUCCESS

Perhaps now is a good time to discuss mentorship with you. Thinking ahead includes choosing and interacting with mentors. No one can advance along a career path alone. You will need support from family, friends, and partners. This core group will provide the foundation, but often they are too close to you and too emotionally involved in your success to be able to advise you

| Use Mentors Wisely |

at arm's length. For guidance, you should first seek out people with knowledge of biomedical, life science, and associated careers, as well as those involved in the general area of career development. They will be able to provide the current best practice information on career building, usually through one or two encounters. Such specific information is important but such interactions, although important, are insufficient.

To succeed, you will need to find individuals in your field who recognize your potential and are interested in your succeeding on your quest. These are mentors. They will advise you on pathways to follow, on opportunities to seek, and on how you should prepare yourself. Best practice is to seek more than one mentor since each will have specific qualities and expertise. Do not get into a situation where your mentors provide you with diametrically opposite advice that will confuse you. Mentors should consider themselves guides and sources of information, but they should leave the final choices to you. Their job is to open your eyes but not to live your life for you. Mentors should not be disappointed with your choices. They should feel proud to provide you with informed choices. Even at an early stage, seek out mentors who can help you with advice on how to deal with barriers that you may need to overcome as you move forward.

## 3.3 THE MENTOR

Over the course of my career, I have mentored many students both at the undergraduate and graduate level. Through trial and error, I learned valuable lessons on effective mentoring. I found that first and foremost, it is best to break down the barrier that exists between the professor and the student; both have to respect each other and interact as friends. The relationship works best if the mentor is not only willing to give but is also willing to learn from the mentee, especially about how students view and react to their environment. So each individual brings something to the table and benefits from the relationship. Once they have gotten to know each other, the mentor should let the students know what areas she can help them with and in which areas they should look for additional mentors. Also the mentor should let them know that she will suggest they see specific people who may be able to help them as specific issues arise. While this is not an exclusive arrangement, if the students regard the mentor as their primary advisor, they should keep the mentor in the loop on major issues, even though these may be out of the mentor's area of expertise. The mentor should request that the mentees meet with her regularly, and inform them that she is available anytime for urgent matters. Also the mentor should make it clear that she expects the mentee to arrive at these meetings on time and to let the mentor know if they are going to be late or must cancel. This is about respecting the mentor's time. It is best to communicate with mentees by email or text messaging, but for confidential matters, it is best to communicate face to face so that no written records go astray.

Mentors should be prepared to submit letters of reference and/or provide phone references. Based on experience, my personal preference is to caution students that I can only comment on what I know about them first hand. Thus if I have never heard them make a presentation, I cannot provide either a written or an oral assessment of their skills in this area. Besides the obvious academic questions, recipients of referee comments frequently ask about the ability of the student to work in groups and to get along with peers and with faculty and staff. They are not interested in hiring someone who will be disruptive to their group. They want to hire people who are helpful to their peers and are willing to share in the work environment. They also want to know about the student's reliability. Does the student arrive at meetings or seminars on time? Does the student let you know if they are going to be late or if they have to cancel an appointment? Questions posed to referees also commonly relate to the student's motivation, creativity, innovation, and willingness to engage in group discussions and projects.

## 3.4 THE MENTEE

What should you look for in a mentee? First and foremost, you should assess the interest the student has in you. This may seem to be a strange consideration since they are the ones looking for advice. However, if the student has not checked you out and does not know who you are, what you do and what your track record is in supervising and mentoring students, then the student may not be serious about developing this mentor–mentee relationship.

As a mentor, you should suggest opportunities for those whom you mentor, in a manner similar to the opportunities that were suggested to you when you were a trainee. It should be made clear to the mentee that this does not mean that the mentee receives preferential treatment. It means that the mentee is encouraged to discuss opportunities that become available, and it is up to them to follow through or not. For example, when one of my mentors was unavailable to teach his group of students, he asked me to substitute for him. I could have looked upon the request as an inconvenience or even as exploitative. It was extra work and I had a full schedule. However, I recognized that he was providing me with an opportunity, so I jumped at it and put a lot of time and energy into teaching well. This led to further requests to teach and to work on curriculum development, which later turned out to be one of my early duties as a faculty member: a task for which I was serendipitously well prepared. In this case, my teaching was a true mentor–mentee collaboration that benefitted both parties.

I have learned that both undergraduate and graduate student mentees overwhelmingly value certain traits in a mentor. When asked to critique their own supervisors and mentors, they universally agreed that academic supervisors—putative mentors—should not only be excellent supervisors but also nurturing mentors. There is a signal difference between supervisor performance, which I discuss later on, and the mentorship relationship that students search for.

In referee letters for faculty teaching/mentorship awards and for promotions, students express esteem for those taking an interest in them beyond the immediate laboratory and research activities. They praise faculty who are accessible to them for career development advice, and who show an interest in helping to advance their careers. However, they show especial esteem for those who provide them with the latitude to explore career options outside the wet bench environment and to try their hand in community activities, including student government, outreach programs to less advantaged communities, and advocacy programs for science.

Some mentors I know go a step further and support their students' extracurricular activities by attending their artistic performances. Generally it seems that students want to have more social contact with faculty, and even want to see their faculty outside classroom and laboratory settings. By helping students build peer relationships with their mentors—as an adjunct to the traditional hierarchal faculty–student relationships that arise from interaction in lectures and laboratories—not only a mentorship relationship but also a tight-knit scientific community with social bonds is created.

The creation of a scientific community, which takes the underlying theory of caring mentoring to another level, also fills a deep-seated need that many students express. The students are looking for the leadership of the department or program to create a nurturing environment wherein the culture includes caring for the students' well-being. They are looking for a collegial community of scholars who will show an interest in what they think about their educational environment and make them feel that they are part of the department. They are not just passing through but want to be a part of the family academically, intellectually, and socially.

A particularly successful way of achieving this goal is through extracurricular athletics, including interdepartmental leagues or faculty versus student sporting events. I fondly remember my own years playing in a noncompetitive baseball league in which departments fielded a co-ed team comprised of faculty, administrative staff, and all levels of students. A natural outgrowth of this culture was the departmental picnic held every summer on Toronto's Center Island, the laboratory beach party/barbecues I attended in San Diego, and gatherings of staff and students in my own home.

The crucial contribution made by the university's administrative staff to the mentorship system should not be overlooked as they provide the infrastructure to run the courses, programs, visiting lecture series, and student committee meetings, and provide the bureaucracy to submit student scholarship and award applications. Students often approach the administrative staff, that is, the departmental undergraduate, graduate, or postdoctoral administrators for advice, both of a personal and professional nature. Students often maintain their friendships with their favorite administrators long after leaving the department to continue on their career paths.

In order to create and maintain a culture of mentorship within both academic and nonacademic institutions and business organizations, evidence-based mentorship training programs are needed. These programs offer a well-defined curriculum and are usually directed toward the organizational leaders who are in the best position to establish a mentorship culture for students, trainees, faculty, and employees. Once the leaders are trained, these programs usually trickle down to middle management in the business world and to faculty at the university who teach undergraduate, graduate, and postdoctoral students in the academic world.

## 3.5 MENTORED RESEARCH AT THE UNDERGRADUATE LEVEL

As an undergraduate student you will continue to be inspired by specific lectures, courses, or programs which will help you focus your research interest in the life sciences and biomedical research. Speaking to professors, teaching assistants, and peers will inform you about research opportunities that you might apply for. Many universities also post undergraduate research opportunities online. Attending undergraduate student research presentations, departmental seminars, and reading the literature will expose you to interesting areas of research. While these experiences sometimes enable you to discover and focus on an area of research you would like to pursue, at this point, learning the fundamentals of best practices in research in a nurturing laboratory environment is far more important than what specific project you do. One of my graduate students told me a few years later that she targeted my laboratory for undergraduate research after she had attended a research poster day in our department. She was looking for an interesting area to work in and spoke to several student presenters to learn about the project and the laboratory environment. When she came for an interview with me, it was obvious that she had checked us out well and knew what she wanted. She was selected as a student, first as a summer undergraduate and then as a graduate student.

> **Undergraduate Research Laboratory Experience is Essential for Future Success**

In order to be accepted to life science graduate programs, undergraduate research experience is crucial. This experience may be obtained in credit courses that feature a research project or through mentored summer research opportunities (often not for credit) that immerse you in a research laboratory doing hands-on work and functioning as a junior graduate student.

Whether or not you have an awesome experience depends not only on your own motivation and hard work, but also on the effective supervision you receive in the laboratory. You should prepare a development plan with your mentor for even a brief 12-week program, so that you achieve the maximum. A convenient way to do so is to create a skills checklist (based on the sample Development Plan provided at the end of this chapter) and build a timetable specifying when you expect to be exposed to and learn the various skills.

## 3.6 BENEFITS OF ENGAGING IN A RESEARCH EXPERIENCE

You will experience and learn good laboratory practice, experimental and critical evaluation skills, collaboration, and communication and networking skills. You will learn about laboratory safety protocols and standard operating procedures. As you do your bench work you will learn to properly document laboratory protocols and results and maintain proper laboratory records. When you begin your research project, you will learn to design experiments, use appropriate laboratory techniques, and be taught how to critically analyze and interpret data. This research will introduce you to troubleshooting problems and actively formulating appropriate solutions, a crucial asset. As a trainee, you will learn both how to search the scientific literature and how to critically evaluate your own scientific data and other scientists' publications. You will learn how to communicate, both orally and in written form, in a scholarly style by participating in seminars, giving poster presentations, including a final presentation of your research work, and by writing reports and journal articles. You will also participate in journal clubs and/or lab meetings. One of the most fundamental skills you need to learn is how to work as a team member in a collaborative research environment, and most but not all laboratories should provide you with this opportunity. As your project evolves, you will also begin to build contacts with peers and professionals within your own research environment and beyond. The latter exposure will introduce you to other perspectives that may shed light on your own research problem.

> **Appreciate Advice on How to Get Better and Fine tune Your Research Skills**

During your time in the laboratory, take advantage of your supervisor, peers, especially senior undergraduate students, graduate and postgraduate students, technicians, and those in neighboring labs to obtain feedback on your work and presentations. At the end of your course, project, or summer experience, thank all those who helped you along the way. A well written thank-you card will be appreciated by those who took the time and made the effort to support you in your research work.

Even at the undergraduate level, scientific societies encourage students to attend their scientific meetings and to present their research, usually in the form of a poster presentation. The societies usually request a

**Undergraduates Should Attend Scientific Meetings**

very low registration fee and they may provide merit and/or needs based scholarships to support travel and board at the meeting. This is an excellent opportunity for undergraduates to meet peers and scientists from other institutions, to hear and see excellent science, and to begin to understand the culture and the community of science. Several of my students attended scientific conferences as undergraduate students and presented their work.

A few years ago, several of my students attended a large scientific meeting and presented their posters. They reported that besides learning new material, they were most impressed and inspired by seeing and hearing presentation from major scientist who they knew only through reading the literature. They were thrilled to speak to researchers whose papers they had read, and their excitement at being interrogated at their posters by a "big name" in their field was palpable. Attending this meeting also gave them the chance to explore possible graduate positions outside of our institution as senior faculty are always on the lookout for high-quality PhD candidates, and at such a meeting you can meet them face to face. This face-to-face meeting may occur serendipitously if a faculty member visits your poster. If they seem to be interested in you and your work, offer to be available for a short private meeting during the conference to discuss your future training plans. If you have identified a few laboratories that you might consider going to, contact the scientists well before the meeting and ask them if they are going to this upcoming meeting, and if you could meet with them briefly to discuss your training. Let them know when and where you are presenting your work at the conference.

Even at this early undergraduate stage, some students may benefit from an international research experience which will broaden their horizons. Should you wish to take such a step, you should solicit advice from faculty, mentors, and peers as well as from your university's international experience office, which will be able to provide you with information on the laboratory and institution you are planning to attend. You should pay careful attention to language requirements and to be sure to honestly assess your ability to adapt to different scientific and personal cultures. As biomedical and life sciences become more global, an international experience can be very valuable to your burgeoning career.

## Undergraduate Research Opportunity Development Plan

Student: Name

       Status

Supervisor: Name

         Rank

         Department(s)

Date Plan Initiated:

Title of Project:

Project Abstract (250 words):

## Skills and Competencies

Check the box if you have been exposed to and/or learned these skills and competencies

☐ General Standard Operating Procedures in Life Sciences Laboratory
   Comment:

☐ Laboratory Health and Safety
   Comment:

☐ Animal Care Course/Procedures (as applicable)
   Comment:

☐ Human Research Ethics Approval (as applicable)
   Comment:

☐ Research Integrity
   Comment:

☐ Code of Student Behavior
   Comment:

☐ Critical Reading of Literature
   Comment:

- ☐ Experimental Design, Analysis
  Comment:
- ☐ Technical Skills
  Comment:
- ☐ Interpretation of Research Findings
  Comment:
- ☐ Oral Communication of Research
  - ☐ To Supervisor
    Comment:
  - ☐ To Laboratory Group
    Comment:
  - ☐ To Fellow Students in the Program
    Comment:
- ☐ Writing Scientific Report
  Comment:
- ☐ Poster Presentation
  - ☐ Local
    Comment:
  - ☐ Elsewhere (Scientific Meeting)
    Comment:
- ☐ Writing Manuscript for Publication
  Comment:
- ☐ Networking
  Comment:
- ☐ Plan to continue this project (Yes; No)
  Comment:
- ☐ My future research training
  Comment:
- ☐ Other
  Comment:

Note:
- Prepare the Development Plan before you start and discuss with your supervisor.
- Fill in the plan as you achieve each competency and complete the full plan at the end of your project.
  In comments, provide some details including outcomes.

## CHAPTER NOTES

_____
_____
_____
_____
_____
_____
_____
_____
_____
_____
_____
_____
_____
_____
_____
_____
_____
_____
_____
_____
_____
_____
_____
_____
_____
_____
_____
_____
_____
_____
_____
_____
_____
_____
_____
_____
_____
_____
_____

# CHAPTER 4

# Choosing the Best Graduate Program for Your Needs

## Contents

## Summary

When selecting a graduate program it is absolutely essential that you be well informed so that the program and supervisor you choose are the right fit for you. The working relationship you have with your supervisor is critical to the success of your studies. Be sure to speak with many students in your prospective programs to get information and opinions about the program and the supervisor. Moreover, ask more than just questions about the curriculum. Find out how the laboratory functions, how the supervisor interacts with her students, and how the department/institution provides resources to enhance your educational experience. Try to decipher information regarding the faculty's level of contentment and the training program's environment. Also take into account personal and family issues when making decisions about graduate programs. These are very important and will impact on the success of your studies. Understand how your research training will progress and be ready to handle failures and rejections along the way. These are par for biomedical and life science research training programs. Understand and take advantage of your student advisory committee which is there to guide and assist you in your program.

*Planning a Career in Biomedical and Life Sciences.*
DOI: http://dx.doi.org/10.1016/B978-0-12-802242-9.00004-9

You need to be well informed when choosing a graduate program since graduate studies is a major commitment of time and effort on your part. Decide which route(s) you wish to fully explore. Do your homework thoroughly and examine several programs so that you can select the best one for you. The fit must be right. If you are lucky enough to have already identified your passion, have fun exploring it. If, like most, you need to investigate several types of research before settling on one to pursue, make sure that you know what you want out of the program you choose and ascertain that you suit the program and that the program suits you.

Speak to as many trainees as possible to get a consensus of opinions about the program. You will find some trainees who are satisfied with everything, and some who can find nothing nice to say about the program. What you need is a balanced assessment, and to get this you will need to ask probing questions and try to get specific answers, not just general impressions. While such an approach is still subjective, if you speak to enough people, you should find out what you need to know. You should also try to get a sense of the faculty's level of contentment, since a happy faculty tends to be much more inclined to create a healthy and exciting training environment. The training program's philosophy and curriculum should facilitate high-quality research opportunities. There should be a culture that is free from harassment and discrimination, and this should be supported by specific policies that are enforced.

## 4.1 QUALITY OF THE PROGRAM

Assessing quality is not a simple matter. The quality of the institution as a whole can be assessed by outcome data and the impact graduates make. These should be available to students. There are also university ranking systems that rank universities based on institutionally supplied data or data derived from third-party sources. Each

Identify the High Quality Programs

approach has its own flaws. Further, you have to be careful of aggregate ranking since it may not reflect the quality of the specific program you are interested in, but only reflect the entire university or institution's standing.

Quality programs often contain diverse student bodies that increase the opportunity for cross-fertilization of ideas by bringing together students from different undergraduate backgrounds.

## 4.2 INFRASTRUCTURE CONSIDERATIONS

- Find out about the financial situation and the research infrastructure at the institution as a whole and in the department you plan to train in.
- Is major equipment readily available?
- Does your potential supervisor have the infrastructure to carry out first-class research?
- Is the research and office space of high quality?
- How good are the library and the associated information technology?

## 4.3 PROGRAM CONSIDERATIONS

- Have core competencies for your program been identified and clearly communicated to you?
- Is there a clear description of the program available to trainees?
- Are there appropriate courses relevant to your research interests?
- Are there opportunities for students to interact and work together?
- Is there a critical mass of trainees in the program and in the institution as a whole?
- Does the program provide opportunities for formal presentations of your student research?
- Is there strong support for student advisory committees and student mentoring? The committee members are crucial because they help keep projects on track, assist in troubleshooting problems that arise, are an excellent source of reference letters, and may provide training in new techniques.

  > **Student Advisory Committees Are Essential**

- Does the program and/or institution have well-defined policies on training and graduate education, including ethical conduct in research, intellectual property guidelines, publication policy, invention policy, safety policy, a code of behavior on academic matters, graduate supervision policy and guidelines, an effective mentorship program and career counseling, and so forth.
- Does the institution/department offer co-curriculum programs/graduate professional skills programs that add to the education you receive in your required degree courses?
- What is the time to completion for the program?

- Are there seminar series and visiting lectureships organized for trainees and faculty?
- Does the department have a cultural climate for innovation and for entrepreneurial commercialization?
- Does the department/institution have opportunities to provide real-work experience for students including paid work term placements?

## 4.4 PERSONAL ISSUES

You should find out whether you have to apply for competitive internal and/or external funding. You should discuss with potential supervisors and program directors numerous issues including salary, benefits, holiday time, conference time, sick time, health insurance, books, computer/laptop, and travel allowances. Some of these are fixed by the institution, department, or program; however, you will be surprised that you do have room to negotiate. Go through each one with your potential supervisor. Often she may be able to add some top up salary or provide you with an additional stipend for some laboratory-related tasks outside of your research training. There are often teaching assistant positions available, but you need to understand how accepting one of these positions will impact your total student stipend and your time in the research laboratory. Make sure you have sufficient time to carry out your research studies for your graduate degree.

Living in an attractive safe neighborhood increases the quality of life for you and your family. Explore housing issues and possible institutional housing. Consider how much time you may have to dedicate to commuting. Institutions should have dedicated individuals who provide information on available services and relocation issues. Involve your family in the decision making. A happy homelife provides the support and peace of mind you will need to dedicate yourself to your training. If both you and your spouse or partner need to train, make sure the institution is able and willing to provide for the academic needs of both members of the couple. Institutions are becoming much more aware of training for double careers and should be willing and able to discuss this with you. It is very important to make the training process enjoyable, so keep this in mind when making your choices.

> **Lifestyle Is Critical**

## 4.5 FAMILY ISSUES

- Does the department or the institution care about you and your family's well-being?

- Does the location allow you to have the quality of life you desire?
- If you have a spouse or partner and a family, make sure their needs are well met. If your partner or family is not happy, you will have to expend precious energy on ameliorating the situation and may even be forced to relocate. Finding a community for your family is important, even if you stay in a location for only a few years.
- Do you need day care or schools?
- What educational programs are available for partners?
- What jobs are available for partners?

## 4.6  INTERNATIONAL STUDIES

You may wish to do a graduate degree at an international location. In fact, certain projects are best undertaken at international locations because of the nature of the work. You may be attracted to a particular institution because of its outstanding reputation, or you may wish to work with a specific scientist whose work you admire. Be well informed. In tandem with a decision to move, you also may decide to strike out in a new research direction. This decision also has its pluses and minuses, so do not take it lightly. You must research your choices very carefully to ensure that you understand the educational characteristics of the institution that you plan to attend and whether you will fit in. There will be differences, often striking ones, between your current institution and department and the international site you are considering.

Discuss your options with a university office dedicated to international experiences and seek out students and faculty who can give you firsthand information and advice on institutions that you are considering. Visiting the prospective institutions is an excellent idea in principle, but is not always feasible. Just remember that while studying in another geographic and cultural environment is very appealing, your main concern is receiving a high-quality education that will further your career aspirations.

## 4.7  SUPERVISOR SELECTION

In many cases, supervisors are not assigned by the program; instead trainees choose from a list of faculty. Meet and get to know your potential supervisor as well as you can. If you are coming from another city or educational institution, take the time to visit. Meet the supervisor, current lab trainees, and staff. See the lab facility and the institution as a whole. The one-on-one relationship with your supervisor is of paramount importance to the success of your graduate training. Personalities are important, so learn about your potential supervisor's personality and make sure it will mesh with yours on both the professional and personal levels. If you have co-supervisors, make sure that the three of you are on the same page. You should not have to function as a referee between your supervisors but as part of a team whose teammates are all pulling in the same direction.

If you apply to a program to work with a specific supervisor who you know from your undergraduate days or by reputation, you still need to do your due diligence. In particular, review the supervisor's curriculum vitae especially publications and grants. The quality and impact of the publications and the size and number of grants will give you an indication of the supervisor's scientific reputation, research interests, and the funds at her disposal to aid you. Many programs allow you to choose a supervisor and a laboratory to train in once you have been accepted to the program. When you begin your training, these programs provide several short electives/rotations in a few research labs that interest you. Having had a positive experience in one of these rotations, you will usually choose that laboratory to do your research thesis in. However, even if this is the case, you should still make sure to research the issues I have mentioned above, as well as those I will list in the next section.

> **The Working Relationship With Your Supervisor Is Important**

## 4.8  A THOROUGH LIST OF QUESTIONS THAT REQUIRE ANSWERS

- How does the supervisor run the laboratory and the research program?
- How much contact time do you want with your supervisor as you train?
- How often are formal meetings held with your supervisor?
- How often will you meet with your student advisory committee?
- Does the supervisor provide constructive feedback in a timely manner?
- Will there be adequate assessment of your progress in course work and research?
- Do students finish their program in a timely fashion (often referred to as "time to completion")?
- Do students publish first-authored high-quality work?
- Is the lab well funded?
- Is there sufficient space and equipment?

- What peer support is there in the laboratory and in the department?
- Are journal clubs part of the laboratory activity?
- Are current students happy with their supervisors?
- Are there visiting scientists presenting seminars and interacting with students?
- Does the supervisor have international colleagues and collaborators?
- Does the supervisor actively mentor students in career development especially with personal development plans?

It is critically important to have an open discussion with a potential supervisor who will answer your questions and provide you with the information you need to make an informed choice. If you are not able to have this discussion, this may not be the place for you to train in.

## 4.9  STUDENT PERCEPTION OF GRADUATE SUPERVISORS

First and foremost, students report that a five-star supervisor will frequently sit down to discuss the progress of your research project with you, providing technical expertise, and gently guiding you through the process, without interfering with your autonomy. Students indicate that they enjoy the latitude of trying out new approaches or ideas in the laboratory. They want to be supported both when they succeed and when they fail at the bench. Many students like their supervisor to actively participate in the wet laboratory. They like to work side by side with their supervisors. In reality, this will happen much more often with a younger faculty member than with a more senior one since the latter will have more non-laboratory duties and activities to attend to in her department, institution, and discipline. When I was a senior investigator, I made sure to pop into the laboratory frequently just to tidy up the wet bench, talk to students at the bench, and look at images through the microscope. I was demonstrating to my students and staff that I was still part of the laboratory team although I spent most of my time on a broad range of laboratory administrative duties.

Students also appreciate a supervisor who is accessible to them outside the framework of formal meetings, so that when a research issue arises they can quickly deal with it. Accessibility is indeed laudatory. Students like quick feedback on their reports and on their applications for studentships, scholarships, and the like. Students also appreciate a supervisor who actively facilitates their contact with other investigators and their laboratory groups, so that they can learn new techniques needed for their own research projects. Students also value the time faculty spend on numerous student advisory committees and on voluntarily applying to judge poster and oral research presentations.

Many students, in reference letters for their own supervisors, express the satisfaction they feel when they are asked to help review a manuscript or grant application that their supervisor has received for review. Students correctly regard this opportunity as a chance to sharpen their critical skills and learn how to write manuscripts and research grants, not as more work being dumped on their plates. They are thrilled that their supervisor, a veteran scientist, is willing to trust them with an important task, usually entrusted only to a mature scientist, and is eager to hear their critical opinions on science which they will then incorporate into their own review. The added value of the exercise, of course, is the time the supervisor takes to discuss the review with them and provide critical feedback before submitting the review which the supervisor will sign off on.

## 4.10 RESEARCH PROJECTS

Before beginning your course work and your research studies, you must choose a research project. Usually this is done in consultation with your supervisor who will discuss several possible projects with you. To help you make your decision you should read the scientific literature pertaining to the research project so that you can make an informed choice. One of our students chose a project on inflammation because as an undergraduate, he realized that inflammation signal pathways were very important in several disease states. Thinking ahead, he realized that this topic presented him with numerous future career options. Another student chose to work on Alzheimer disease because she watched one of her closest relatives battle this disease. This experience sparked her desire to understand the condition better and to contribute to the development of possible therapies. While personal interest in, and even passion for, the research topic is crucial, in order to ensure your supervisor's full attention and support, your project should be one that your supervisor is also very interested in and that fits in well with the overall program of the laboratory; a happy circumstance that will occur if, as I have mentioned above, you choose a supervisor who is an expert in your field of interest. Of course, the project should also be challenging enough to presage the likelihood of your completing innovative work publishable in a high-impact journal.

## 4.11 BUILDING YOUR CONFIDENCE

As a graduate student, you will recognize that your skills, maturity, and confidence evolve over time. Initially graduate training can be an overwhelming experience. You must successfully jump several hurdles almost simultaneously. You must define or stake out your space in your new laboratory and office; you must find your way around the institution; and you must begin to learn protocols and techniques and determine how the lab runs. First interactions with your lab mates may be difficult. Identify which peers can help you and mentor you in these early days. Do not be shy to ask questions. Very early on, you must develop sound time management skills and learn to multi-task efficiently and effectively.

Even though you will suffer many setbacks as you probe deeper into your area of research, you must not despair. This is par for the course. Every scientist experiences such frustrations. Just remain focused on your goal and continue searching for a solution to your biological question. By staying the course, you may finally arrive at an awe-inspiring payoff. The excitement of discovering something new, be it a new gene or protein, a new cellular metabolic or signaling pathway, or new insight into cell function and/or how a disease progresses is almost indescribable. Even more exciting is discovering that a previous conceptualization was wrong and you now have the scientific evidence to prove it. (Think how Copernicus, Galileo, and Einstein must have felt! You too can join this pantheon, albeit probably in some more minor way.) Several of my colleagues have had the good fortune to be the ones to finally identify something that others have worked on for years. The persistence of the collective scientific community finally paid off!

## 4.12 INSIGHT

Scientists often gain insight into a problem that they are dealing with in the laboratory while carrying out some unrelated tasks at home or on vacation. As long as a problem is embedded in your unconscious mind, sudden "aha" or eureka moments may occur in the strangest places. This has led many to note that insight is often the product of a prepared mind. Beginning with the tale of Archimedes in his bathtub and continuing with Sir Isaac Newton under his apple tree, science has a long mythology of such insight. More recently, Sir Alexander Fleming's discovery of penicillin, Henri Becquerel's discovery of spontaneous radioactivity, and Percy

Spencer's invention of the microwave oven are good examples of such serendipity. Stepping out of the lab may also be valuable because it enables us to look at the same old problem in a fresh way, to think outside the box: another daring method of problem solving.

The challenge for scientists then is to design clever experiments to reveal the secrets of biological processes, in effect forcing nature to divulge its secrets. Scientists spend a considerable amount of time reading and thinking about a biological question before deciding on how to test a hypothesis that will provide new knowledge. What can be more daunting and at the same time, more fun than identifying a biological question that you find exciting? You need to know the answer, and as an active scientist you have the opportunity to design and plan experiments that will try to explain the biology that so fascinates you. You get to carry out the investigation. You decide what techniques you need to use and learn how to use them correctly and safely. Then you carry out the experiments, troubleshooting any technical issues that bar your way. To make sure no errors have crept in, you repeat the experiments several times until you are certain all is well. Then you carefully analyze the data using the best qualitative and quantitative methods available to you. Once you are confident of the validity of the data, you interpret your findings. Have you discovered something new? Can you explain what you have found in terms of other experiments published in the literature? Most often even success leaves you asking new questions, as your journey to discover the answer to your question takes you to even deeper levels of understanding.

## 4.13 PEOPLE SKILLS

This skill set is especially important in life science and biomedical research primarily because the research is carried out within a scientific community. Your community is first and foremost your own laboratory. Some labs are small, comprised of only a few people (I began in a lab comprised of just myself and one technician), and some are large, comprised of several undergraduate and graduate students, postdoctoral students, research associates, technicians, and staff. You have to be able to function efficiently and effectively within the lab setting. You may have to share lab bench space, equipment, imaging stations, and so forth, and you need to learn how to be firm and accommodating at the same time. Be a good citizen of the lab. Do not book time on a piece of equipment for several days, effectively preventing your colleagues and peers from completing their own work.

Discuss with other users of the equipment how best to satisfy every-one's needs. If you realize that you will not be using equipment that you booked, immediately remove your name from the time sheet and let colleagues know so that they may take advantage of this free time. Clean up after you use the equipment and replace any supplies so that the next person does not have to scramble to get supplies just as they are about to start using the equipment. Report any malfunctions immediately so that they may be fixed and others are not inconvenienced. If you are using common equipment or equipment in a colleague's laboratory, do not overstay your welcome and make sure you know how to use the equipment so that you don't take it out of service. Sometimes you may have to stay late or come in early to use equipment in high demand, however, help create a culture in which these early or late time slots are distributed amicably and equally among the group of users.

On the personal side, say "good morning" or "have a good evening/weekend." Smile and don't be grumpy. Say please and thank you, especially when others do you a favor. Sometimes a colleague needs a helping hand in navigating a particular experiment, so be ready and willing to lend a hand. The lab environment is a close contact environment so that being courteous and considerate is a *sine qua non* for making life in the science laboratory pleasant and enjoyable for all.

## 4.14 RECORD KEEPING AND RESEARCH APPROVALS

As you work on the bench, you must be aware of ethical issues in research, show that you are carrying out the best scientific procedures and follow the highest ethical research standards. To put it bluntly, anyone should be able to read your records and published work, understand and approve of your scientific process, and duplicate your experiments. Falsification of data or research outcomes is a heinous crime, and data preservation (in the form of lab books) is central to preventing ethical misconduct and to keeping to the highest ethical standards.

Therefore, you must learn to keep a lab book chronicling your daily activities in the laboratory. Some labs are turning to homegrown or commercially available electronic lab books. Experiments should be meticulously described, and data should be scrupulously kept. All entries must be dated. The page numbers of the lab books must be inscribed consecutively. The book should be kept in a safe place in the laboratory so entries can be made in real time and the book does not get lost or misplaced in transit out of the lab. Original data, images, and data analysis must be kept in a well-organized fashion so that they are readily available for review or for audit purposes. Every image or graph you keep for your records should be of publication quality. All your original data should be encrypted so that it is secure and if lost, remains secure. If you are privy to any data or information that involves human beings, ensure your familiarity with institutional rules and guidelines for handling human research information, including requirements for human research ethics board approvals.

## 4.15 SUPERVISOR MEETINGS

Initially you may find that meeting with your supervisor to discuss your work is intimidating. In order to get the most out of a meeting, first send an email with your data to your supervisor ahead of time, so that she will have time to think about your work before you meet. Second, prepare carefully for the meeting. Consider what message you want to convey and what help you need to ask for. Be open and address issues in a clear comprehensive way showing all the data, the good and the not so good. Don't use the phrase "you are not going to like the data," because this implies annoyance that the data does not fit the hypothesis and presumes that your supervisor is going to be unhappy. Your supervisor should not be perturbed by this; as long as your experiments are well designed, carried out in a competent fashion, analyzed, and interpreted, your supervisor should be pleased.

In discussing your methods and outcomes with your supervisor, you must take responsibility for your project—after all, it's your project. Be confident of your methods and outcomes. No one knows as much as you do about your project. Just remember that everyone needs input and advice.

Once you begin your research you must discuss authorship and the institution's policies on intellectual property with your supervisor. You should understand what intellectual input and work is required to claim first authorship, which in life sciences usually means you are senior author. If you receive senior author billing then your supervisor becomes senior responsible author usually referred to as corresponding author. Disagreements about authorship most often arise when more than one graduate student or more than one principal investigator are involved in a project. Then a plan for authorship should be committed to writing at the beginning of the process to preclude arguments and misunderstandings that might come up later. The original plan may change as the research progresses, but any adjustments should be made to everyone's satisfaction. Sometimes you may need a neutral party, such as a graduate coordinator, to help resolve authorship issues.

## 4.16 STUDENT ADVISORY COMMITTEE

In most programs the student advisory committee is a requirement of the program, and if it is not, then it should be. There are departmental/institutional guidelines to inform the student, supervisor, and committee members on the role of the committee, how to conduct the meetings, and what to expect from them. The meetings should be held approximately every eight months. The committee is comprised of the supervisor and two or three faculty members. The department may require that one member be from another department. The committee members should be chosen by the student in consultation with the supervisor and they should be free of conflict. The members should have the scientific expertise to provide the student with advice and to assess his or her progress. While the committee does evaluate the student's work to assess progress and determine if the student is achieving research competencies, the meeting is not strictly speaking an evaluation process.

Usually a typical meeting will begin with a presentation by the student emphasizing experimental design, execution of experiments, data analysis, and conclusions. The student should prepare a full report for distribution to the supervisor and committee members 7–10 days before the meeting. The report should be well written, so the student should consult with peers and faculty on the composition of the report and take the time to write several drafts. The committee meeting is the time and place for the student to put his or her best foot forward in a protected environment. The presentation will be followed by a general roundtable discussion that should focus on the presentation and report. The supervisor should allow the student to engage in the full discussion. When it is necessary and appropriate, the supervisor should contribute to the discussion.

Sometimes the committee members and the supervisor may get involved in cross-talk. This is not helpful to you. You need to ask your committee members questions and discuss conceptual and technical issues that are on your mind. It is important as well to present and discuss experiments that did not work. To your surprise, you may get useful feedback on the failures as well. You may also find that often the committee members are willing to have you come to their labs to refine and/or learn new techniques that will aid you in your research.

A report of the meeting, often based on a template created by the department or program, is usually composed and signed by all parties. Helpful comments on the research, suggestions on the direction of the work, and the goals and outcomes to be achieved by the next meeting should be clearly recorded in the report and agreed upon by you, your supervisor, and the other committee members.

The overall tone of the student advisory committee meeting should be collegial and supportive of your efforts, but this also means an honest critique of your efforts will be provided. Since the emphasis should be on guiding your work, you must provide the members with all your work and have the raw data available, especially if there are questions about the data analysis. If your progress is lagging this should be made clear to you, and suggestions on how to improve your pace should be included in the report. You must monitor time to completion of your thesis work, and the committee should provide the direction necessary to help you keep an appropriate pace.

The committee members in consultation with your supervisor will determine what constitutes an appropriate thesis for you. Make sure you understand the expectations. When you have completed your research work, the committee will grant you permission to write up your thesis, and once you have done so, you will defend your work at your thesis defense. Be clear on what the thesis defense consists of in your department and university. Usually there is an established process which you must be aware of. The earlier you understand the process, no later than a year before completion, the easier it will be to prepare the thesis and to prepare yourself for the defense.

Committee members should be available to you outside of the meeting format for advice between meetings. Some students find a one-on-one meeting a very useful addition to the process. A committee member may also be able to mentor you as well, although many students prefer the two tasks be kept separate. As your project progresses, your supervisor and yourself may wish to add another member to the committee, usually to provide additional scientific expertise. On occasion you may find that a member is no longer useful as your project moves forward. You should discuss this with your supervisor and perhaps make the necessary change. Your committee members will get to know you so they may be fine sources of letters of reference for awards and scholarship applications, and for postdoctoral and job applications.

## 4.17 LITERATURE

Read the professional literature extensively, critically, and in a timely fashion. You cannot possibly carry out and interpret your own research work

| Be Current with the Literature |

or broaden your general biomedical and life science knowledge base without doing so. By the same token, always stay current and be ready to discuss the latest breakthroughs in your field and in science in general.

## 4.18 COMMUNICATION

You must learn how to write scientific articles and present scientific talks. Some institutions provide face-to-face and/or online modules to train students. While in graduate school, pay attention to the "soft skills" and make sure that you communicate well, both orally and in writing and that you learn to network and to communicate with peers and colleagues at scientific meetings. The latter skill is crucial because it allows you to meet scientists in your field and begin to build your scientific reputation.

But what happens if you happen to find yourself walking across campus next to one of your scientific idols or sit down next to one of them on a public bus or at a conference? What should you say to them? An important skill that you will often hear spoken of in marketing classes is the "elevator pitch." Analogous to "speed dating" in the social scene and the "three minute PhD" in the academy, this term refers to a scenario in which you are riding in an elevator with someone and you have about 30 seconds to answer the question she poses to you, "What exactly is it, that you do?" What should you say to your scientific idol to engage her interest? The scenario is meant to force you to develop an honest, brief, and comprehensive response to someone whom you need to impress in a painfully brief amount of time. Thus you have to be able to communicate key information in a succinct and unambiguous manner that highlights the main points rather than getting bogged down in the details.

One way of developing your scientific "elevator pitch" is to use a template approach that begins with the reason you are doing this type of research project; the hypothesis you are testing; the methods you use; the new knowledge that you generated or hope to generate and the importance of the discovery to the listener and to the community as a whole. Your goal is to have your listener recognize you as a confident and interesting individual, who is performing important research. The typical listener will ask herself "What does this mean to me?"; "How does your work relate to my interests?" If you are very successful your listeners will ask for more information and perhaps request your business card so that they can communicate with you further at their convenience.

It is very hard to come up with an elevator pitch on the spot. You need to prepare the pitch, refine it, and rehearse it numerous times, so that when the opportunity arises you can deliver it flawlessly. As you become more experienced, you may choose to spontaneously tailor a sentence to hone in on a particular interest of the listener. The pitch is important when you speak to potential postdoctoral supervisors, academic and nonacademic recruiters, potential collaborators, network buddies, research funders, including those from nongovernmental agencies, those from the world of commerce and industry, and philanthropists. The pitch is designed to help you get your foot in the door. If you have gotten your listener's attention, you will have the opportunity to provide much more detail later on.

> **Prepare business cards for yourself**

## 4.19 FAILURE

While failure pervades scientific investigations, the reasons for particular failures should be investigated so that they can be prevented if possible. Failures can be linked to several different factors: (i) Your idea or hypothesis is fundamentally unsound. It is based on an incorrect interpretation of the literature, or it is poorly conceived to move the science forward. (ii) Your technical skills are not of the highest quality, so your data is meaningless. (iii) You cannot carry out your experiments in a satisfactory matter because of experimental design flaws or technical problems. For example, in tissue culture your cells do not tolerate the protocols you wish to put them through although you may have made numerous modifications. You will ultimately have to either give up the project completely because the technology necessary is unavailable or adopt the more likely scenario of redesigning your experiments. (iv) You may misinterpret your findings and move forward in an inappropriate manner to the next set of experiments.

Failures cannot be avoided but you can reduce their prevalence by discussing your hypothesis and planned experiments with supervisors, colleagues, and technical staff, and by presenting frequently to your lab group to get useful feedback. To avoid technical failure and add more skills to your toolkit become proficient in specific technical skills, especially those not being used in your lab. To do so, in addition to reading the literature, you may need to seek advice from those with experience using the technique. Visit a lab that has the technique up and running successfully. These experts can provide you with tips that you will never read about in any materials and methods descriptions in a published paper. Consulting in this manner will prevent you from wasting precious time and resources as you carry out your experiments.

The most difficult failure to contemplate is leaving your program without completing it and not obtaining your degree. If this occurs very early on in your program then you end up losing some time, but you gain much more as you can immediately move on from a program that you have lost interest in and that you no longer find enjoyable and fun. Rather than considering this a failure, think of it as a well-thought out reorientation of your interests and career path. However, if such a failure occurs late in the program, your natural instinct may be to attribute this to a failure to perform and you may be quite discouraged. To avoid the latter scenario, do not let your feelings smolder. As soon as you feel a loss of motivation, speak frankly with your supervisor, graduate coordinator, and/or mentor. This may help you quickly regain your resolve.

Often graduate students have very high expectations of success and expect almost immediate results. Unfortunately science does not work this way. Instead you will be frustrated by many technical problems and poor experimental design, and you will experience many failed experiments and hard to explain outcomes. If you become disillusioned at the poor return on all your hard work, learning and understanding how research projects evolve and that your experiences are not unique may improve your mood. The difficulties in producing good science are legion and you will need to learn to live with them.

Speaking to faculty may also uncover other problems that if corrected, would cease to hamper you, so that you could get on track once again. However, if you are truly unhappy and are rapidly losing interest, then you must make an informed choice. Stick it out and finish your degree as best you can or cut your perceived losses and resign from your program. It is far better to leave gracefully then to be terminated due to consistent failure to meet program standards. If you have been making careful informed decisions throughout your training, you probably will not find yourself in this predicament. However, there is no foolproof way of preventing such a situation from occurring.

Sometimes you may feel that your supervisor, your department, or your program is holding you back. While you may switch to another supervisor, department, or program if all the necessary conditions are met, you need to carefully consider whether these are the real causes of your dissatisfaction. While switching to a new supervisor or project may appear to be the panaceas you are looking for, the same issues may arise again. The grass may not be greener on the other side. You need to seriously consider whether a change in scenery will resolve your difficulties or whether it is best for you to take a break and/or consider a different pathway. If you decide to switch to another supervisor or department, you must speak to your graduate coordinator (or equivalent) to seek advice on whether this is an appropriate option for you and if so, how to set the process in motion. Be aware of the time you will lose starting again and make sure that there is student stipend funding in place in your new setting.

## 4.20 REJECTION

Rejection is always hard to handle, but the true scientist must show his or her mettle by bouncing back from rejection (or failure); resilience is the key. It will be easier to bounce back if you understand that rejection is part of the "game"; the playing field in the scientific arena can sometimes be harsh. In life sciences the two important areas of rejection are manuscript and grant-in-aid rejection. If you have submitted a manuscript or a research grant and it has been rejected by a journal or a funding agency, the journal or grant reviewers will provide reasons for the rejection. You have to learn how to read these critiques so that you do not become overly depressed. Often with the help of mentors and colleagues, you may come to realize that the paper or grant just needs minor changes, not major modifications. Sometimes you may need to perform additional experiments, which could take up to a year or more to complete, but do not be discouraged. Learn from the experience. The review process is designed to make your science better. However, do consult with others to make sure that there are no fatal flaws in your manuscript or proposal which cannot be overcome by modifications. If this is the case, you may need to make major changes that are tantamount to actually abandoning the project and starting over.

Remember that decisions on grants are not normally appealed; however, you can still register a thoughtful complaint with the granting body to express your disagreement. Likewise, if you feel you were unjustly treated by the reviewers of your submitted manuscript, you may communicate this to the journal editor. However, the editor has the final say and will let you know if she feels your complaints merit further examination. As a resilient scientist, you must learn from these setbacks and heed the reviews, so that you can march successfully onward. The key to success is carefully and soberly listening to the criticism so that you can benefit from it. In my own experience, my lab was carrying out *in vitro* studies in a model system and we concluded that our studies shed useful light on the more complex *in vivo* condition. This interpretation was criticized by both grant and journal reviewers. Once we heeded the critiques and carried out a series of *in vivo* experimental studies to validate our *in vitro* findings, we were able to continue our work in the simpler *in vitro* systems to direct our studies at biological mechanisms of disease.

## 4.21  SATISFACTION IN TRANSLATIONAL RESEARCH

Many life scientists want to take their basic science discoveries and translate them into products or treatments that have social and medical benefits either for individuals or entire communities. These applications may treat a disease or be directed at a communal or global scientific problem such as global nutrition or child and women's health. Those who wish to take this path often refer to it as "the bench-to-bedside" approach to translational research; the translation of their basic discovery in the experimental laboratory, into a diagnostic, prevention, or treatment tool. After successfully completing clinical trials in humans, this tool may enter regular clinical practice. The term translational research may also be used to describe the promotion of a known product into general use within the community, or to describe research performed on clinical material obtained from patients in order to develop potential diagnostic, therapeutic, and/or preventative strategies in disease conditions.

Although graduate research spans a relatively brief period of time and is carried out by relative newcomers to the field, some trainees are able to initiate the bench-to-bedside pathway by discovering new knowledge that may very well produce novel treatments or a new drug or device that improves health care. In some cases serendipity leads to the innovative finding, but in others the student and supervisor spend considerable time developing a research plan designed to discover new knowledge that has serious potential for innovation. In the latter case, the student and supervisor will have arrived at a well-developed research question and will have made sure that the necessary resources and technology are available to carry the entrepreneurial project forward. They will also have determined societal need for their product, should the research be successful, and considered possible commercialization pathways and investors. Thus, they have a very good chance of successfully launching into product development. Depending on how long the discovery research takes, there may be time to carry out some initial steps in product development during the PhD studies and thesis work. Otherwise this will have to be carried out after the degree is granted.

Once the student receives her degree, she has to make a choice: Does she want to continue developing the finding along a translational paradigm, or does she want to hand it off at this point and move on with her own life science training? Alternatively, she may choose some form of minimal involvement. Whatever course she takes, she must review and clarify the intellectual property issues with the other parties involved. If she decides to continue developing her finding, she will have to acquire new information and skill sets, as well as mentors familiar with product development and commercialization. It will usually take several years to fully complete the process. Universities and research institutes continue to develop infrastructures to assist students in these endeavors, creating effective innovation ecosystems to advance the translation and commercialization agenda, especially in the early stages.

I am familiar with numerous "bench-to-bedside" discoveries made by students in my own department. One of our PhD students identified tumor-specific biomarkers that distinguished between two forms of a cancer, a discovery that is leading to promising drug therapy strategies. Another student made a new discovery using animal models that has the distinct possibility of developing treatments that may prevent blindness. These achievements are inspiring to graduate, undergraduate, and postdoctoral students working in biomedical and life sciences.

## Graduate Development Plan

Student: Name

       Status

Supervisor: Name

         Rank

         Department(s)

Date Plan Initiated:

Date(s) Plan Reviewed/Updated:


Title of Research Project:

Project Abstract (250 words):


## Courses Required and Completed:


### *Learning Procedures and Competencies*

*Check the box if you have learned this skill/competency.*

☐ General Standard Operating Procedures in Life Science Research Laboratory

   Comment:

☐ Laboratory Occupational Health and Safety

   Comment:

☐ Animal Care Course/Procedures (as applicable)

   Comment:

☐ Human Research (as applicable)

   Comment:

☐ Radiation Protection Course (as applicable)

   Comment:

☐ Research Integrity/Ethics

   Comment:

☐ Code of Student Behavior

   Comment:

### *Research*
- ☐ Critical Reading of Literature
  Comment:
- ☐ Experimental Design, Analysis
  Comment:
- ☐ Technical Skills (List)
  Comment:
- ☐ Interpretation of Research Findings
  Comment:
- ☐ Teamwork
  Comment:

### *Communication*
- ☐ Oral Communication of Science
  - ☐ To Supervisor
    Comment:
  - ☐ To Laboratory Group
    Comment:
  - ☐ To Fellow Students in the Program
    Comment:
- ☐ Poster Presentation
  - ☐ Local
    Comment:
  - ☐ Scientific Meeting
    Comment:
- ☐ Writing Scientific Report
  Comment:
- ☐ Networking
  Comment:
- ☐ Writing Manuscript(s) for Publication
  Comment:
- ☐ Other
  Comment:

### Teaching/Mentoring:
- ☐ Teaching Assistant
  Comment:
- ☐ Mentoring Undergraduate/Graduate Students
  Comment:
- ☐ People Skills
  Comment:

## Milestones (include expected date and when achieved)

- ☐ Student Advisory Committee Members (List with brief comment on expertise.)
  Comment:
  Date of Committee Meetings (every eight months)
  1.
  2.
  3.
- ☐ Seminar Presentations
    - ☐ Local Peers
      Comment:
    - ☐ Department/Institute
      Comment:
    - ☐ Other
      Comment:
- ☐ Comprehensive Exam (as applicable)
  Comment:
- ☐ Submission of Manuscripts
  Comment:
- ☐ Publication of Manuscripts
  Comment:
- ☐ Permission to Write up Thesis
  Comment:
- ☐ Thesis Defense
  Comment:

## Noncurricular and Extracurricular Activities:

- ☐ Departmental Student Groups
  Comment:
- ☐ University/Institute Student Groups
  Comment:
- ☐ Outreach Programs
  Comment:
- ☐ Personal Activities at University or Elsewhere
    - ☐ Sports
      Comment:
    - ☐ Arts/Music
      Comment:
    - ☐ Hobbies
      Comment:
    - ☐ Other
      Comment:

Note:
- Prepare the Development Plan together with your supervisor before you start.
- Review and amend the plan at the end of every eight months.
- In comments, provide some details and identify specific outcomes that you have achieved.

## CHAPTER NOTES

_____

_____

_____

_____

_____

_____

_____

_____

_____

_____

_____

_____

_____

_____

_____

_____

_____

_____

_____

_____

_____

_____

_____

_____

_____

_____

_____

_____

_____

_____

_____

# CHAPTER 5

# Postgraduate Studies: Preparing to Launch

## Contents

## Summary

The search for postdoctoral positions can be a long process; therefore, you should begin your search well in advance of completing your PhD, at least a year to 18 months beforehand. Similar to selecting a graduate program, you must take careful consideration in selecting a postdoctoral position. You need a very good fit with your supervisor and a productive laboratory. Postdoctoral education will allow you to learn new techniques and disciplines which will enhance your current knowledge base. Furthermore, it will allow you to carry out research with very little supervision and produce high-quality research that you can present at scientific meetings and publish in high-impact journals. This will allow you to network with others in your field. Be sure to learn how to manage and operate a research laboratory and how to review scientific manuscripts and research grants while you are still in your postdoctoral training environment. This is important as you will be required to perform these tasks when you have your own laboratory. As a postdoctoral trainee, work on improving communication skills and build your scientific networks. Pay close attention to your progress in the research laboratory since you have limited time to produce high-impact work.

## 5.1 CHOOSING A POSTDOCTORAL POSITION

You should have been exploring postdoctoral positions well in advance of completing your PhD training. The search can take more than a year from preparation of application to landing the position. Popular labs fill up quickly and there may be waiting lists. Many positions are advertised in the fall, with interviews taking place in January to March. Initial contacts can be made in a variety of ways: at scientific meetings, visiting lecturers, job postings in journals and web sites, word of mouth and networking, and through other informal means. Your supervisor and other department members can be very helpful as well. Even if you are applying for a position at your current institution or at one you know well, treat it like an unknown institute because you are now looking at it through the lens of a postdoctoral applicant.

*Planning a Career in Biomedical and Life Sciences.*
DOI: http://dx.doi.org/10.1016/B978-0-12-802242-9.00005-0
53

Postdoctoral training is almost universally required if you plan to continue in academia. It is also useful if you enter the world of industry, business, and commerce. Do your postdoctorate in the best place you can find (and be accepted to), preferably in an institute where several investigators in your field interact and collaborate to form a strong research group. You may wish to interview with more than one investigator in this group to identify the best supervisor for your needs. As you did in choosing a graduate program, you must also make sure that the location will provide a fine quality of life for you and your family (see Sections 4.4 and 4.5).

## 5.2 POSTDOCTORAL CAREER PLAN

Templates are now available for developing a Postdoctoral Development Plan (see below). You should complete such a plan with your supervisor/advisor's input, and use it as a dynamic guide to chart the course of your postdoctoral training. Create milestones along your path to completion that will help keep you focused on the endgame in the open-ended environment most postdoctoral trainees work in. Paying attention to completion is crucial. Revisit the plan and the milestones frequently to make sure that you are on track. Setting these goals and outcomes with milestones is critical for a successful postdoctorate. In many jurisdictions, institutions are applying limits on the time you may spend as a postdoctoral trainee, now generally three years with the possibility to extend for up to an additional three years, if deemed necessary for academic reasons. Thus you need to pay attention to your progress in order to have completed a significant body of high-quality work during this time.

## 5.3 POSTDOCTORAL OUTCOMES

Biomedical and life sciences provide you, as a postdoc, with three important opportunities. First, you may learn new techniques and disciplines which will enhance your knowledge base. Second, you will have the opportunity to carry out independent research with very little supervision. While you are still in a protected research

| Showing Independence |

environment, you can take ownership of your project and be highly innovative on your own. This is the ideal training for a budding independent investigator. Third, you should be able to produce excellent research which you will be able to present at scientific meetings and which you should leverage so that you meet other investigators in your field at these meetings. You should also apply for presentation awards offered at these meetings. At best, your postdoctoral research will result in high-impact first-authored publications. These are critical to competing successfully for academic positions, especially at research intensive universities. During the latter stages of your postdoctoral studies, some institutions may encourage you to apply for postdoctorate-to-faculty transition grants, if available and suitable. These are extremely useful in facilitating your transition to a faculty position.

## 5.4 LEARN MANAGEMENT SKILLS AND THE BUSINESS OF SCIENCE

As you carry out your postdoctoral research, you should be actively learning how to operate and manage a laboratory. Learn the essential components of these tasks and feel free to ask clarificatory questions of lab managers, supervisors, and senior scientists. No formal course is usually provided to teach you these skills, so apprenticeship and keeping your eyes wide-open are the best ways to learn them. In a large laboratory, management may be provided by laboratory staff, so you may overlook its importance; however, once you are required to build and manage a science team on your own, you will regret having not paid attention during your own training. In the near future, the graduate and/or postgraduate curriculum may, indeed, include formal instruction on laboratory management including financial planning, budgeting, human resources, team building, and so on and so forth.

> **Learn the Business of Science**

## 5.5 UNDERSTANDING MANUSCRIPT AND GRANT REVIEW

If you have not had an opportunity to review manuscripts or grants as a graduate student, you must learn how to review scientific manuscripts and research grant applications now, so ask your supervisor if you can help her with those tasks. Learning how to critique others' work also teaches you how to critique and improve your own.

Your supervisor can offer you the templates that different journals use to guide reviewers in performing reviews. These will give you an idea of what the editor and the editorial board are looking for in a manuscript. Each journal has a statement describing the types of studies they publish, that is, a statement delineating the field or discipline of study and/or the techniques their readership is interested in. This defines the aim and the scope of the journal. For example, there are many journals that publish in the area of cardiovascular science and medicine. Some may focus on the heart, some on blood vessels, and some on both. Some focus on basic science or translational and/or clinical studies. Some may be disease oriented, publishing on atherosclerosis, lipids, or transplantation. Others may focus on biomedical conditions such as heart failure or congenital heart disease. You will learn that one of the most important criteria for having your manuscript accepted for publication is choosing the most appropriate journal for it. Even an excellent study will be summarily rejected if it does not fit a journal's scope.

In writing reviews, you will also learn the criteria that editors ask reviewers to rank the manuscript on, such as originality and quality of experimental design, execution, and data analysis. The editor will request an appraisal of the statistical analysis as well as the quality and usefulness of the figures and tables. Last but not least, the editor will want to know if the manuscript contains new knowledge and if it is likely to impact the field and science in general. If the manuscript simply contains a rehash of work already in the literature, perhaps using a different model, it is much less likely to be published in a quality journal. So in order to review other's work properly, you have to be familiar with the literature on the topic. Once your review is complete you are usually asked to assess for publication: Accept as is, Minor Modifications, Major Modifications, Reject with chance to resubmit, or Reject.

Having completed the review, you will compare notes with your supervisor, and she will sign off on the review and send it to the journal editor. If there is any suspicion of breach of scientific integrity, you must communicate this to the editor in the confidential "Response to the Editor." During this process, it will be impressed upon you that the entire process is confidential and you may not utilize or communicate the contents of the manuscript to anyone. The author(s) will not know who you are since you do not sign the "Comments to the Author." Your anonymous "Comments to the Author" is submitted to describe your findings and suggestions about improving the manuscript. If you are proposing modifications or a possible resubmission, the author(s) needs to know and understand what your scientific concerns are about the manuscript. Many journals will send all the "Comments to Authors" to every one of the manuscript's reviewers (usually there are two or three) so you may compare your review with others. The journal does this solely as a courtesy, but as a postdoctoral student you will benefit immensely from this part of the process as well.

## Postdoctoral Development Plan

Student: Name
        Status
Supervisor: Name
          Rank
          Department(s)
Date Plan Initiated:
Date(s) Plan Reviewed/Updated:

Title of Project:
Project Abstract:

## Learning of Skills and Competencies

☐ General Standard Operating Procedures in Life Sciences Research Laboratory
   Comment:
☐ Laboratory Health and Safety
   Comment:
☐ Animal Care Course/Procedures (as applicable)
   Comment:
☐ Human Research Ethics Approval (as applicable)
   Comment:
☐ Research Integrity/Ethics
   Comment:
☐ Code of Student Behavior
   Comment:
☐ Critical Reading of Literature
   Comment:
☐ Experimental Design, Analysis
   Comment:
☐ Technical Skills (List)
   Comment:
☐ Interpretation of Research Findings
   Comment:
☐ Oral Communication of Science
   ☐ To Supervisor
      Comment:
   ☐ To Laboratory Group
      Comment:

☐ To Fellow Students in the Program
Comment:
☐ Poster Presentation
    ☐ Local
    Comment:
    ☐ Elsewhere (Scientific Meeting)
    Comment:
☐ Writing Scientific Report
Comment:
☐ Writing Manuscript(s) for Publication
Comment:
☐ Other
Comment:

## Teaching/Mentoring:
☐ Teaching Assistant
Comment:
☐ Mentoring Undergraduate/Graduate Students
Comment:

## Milestones (include expected date and when achieved)
☐ Student Advisory Committee Members (List with brief comments on expertise) (if applicable)
Comment:
Date of Committee Meetings (every eight months)
1.
2.
3.

## Noncurricular and Extracurricular Activities:
☐ Departmental Student Group
Comment:
☐ University/Institute Student Group
Comment:
☐ Outreach Programs
Comment:
☐ Career Development Seminars/Courses/Meetings
Comment:
☐ Career Networking
Comment:
☐ Seminar Presentations
    ☐ Local Peers
    Comment:

- ☐ Department/Institute
  Comment:
- ☐ Other
  Comment:
☐ Submission of Manuscripts
Comment:
☐ Publication of Manuscripts
Comment:
☐ Aid supervisor in Manuscript/Grant Reviews
Comment:
☐ I plan to continue my project after my postdoctoral training.
Comment:

Note:
- You have learned several of these skills as a PhD student. Comment on your level of competence and whether you need additional training.
- Prepare the Development Plan before you start, together with your supervisor.
- Review and amend the plan at the end of every six months.
- In comments, provide some details and identify specific outcomes that you have achieved.

## Skill Sets/Competencies You Achieve by the End of Your Formal Training

There are numerous skill sets and competencies you should have attained during your training period to promote your success. Your attainment of these skills will be incremental as you move through your training. Some will be achieved in the course of your work and others will require specific efforts by yourself to find out where to learn these skills and from whom.

### Research:
- **Research Methods, Design, Data Analysis, Critical Analysis**
- **Large Data Organization, Analysis, Presentation**
- **Technical Expertise**
- **Inter/Transdisciplinary Research**

Education:
- **Self-learning (Life-long learning)**
- **Teaching Skills**

Communication:
- **Team Work/Collaboration**
- **Writing**
- **Presentations**
- **Social Media**
- **Networking**

Management:
- **Laboratory Management**
- **Small Business Skills including Negotiations, Human Resources, Financial Administration**
- **Strategic Planning**

Community:
- **National and International Scientific Societies**
- **Science Advocacy**

# CHAPTER NOTES

_____
_____
_____
_____
_____
_____
_____
_____
_____
_____
_____
_____
_____
_____
_____
_____
_____

# CHAPTER 6

# Your First Job: Choosing and Preparing Well for an Academic Career

## Contents

## Summary

Similar to applying for postdoctoral positions, the job search process is also long and you should begin sending out applications at least one year in advance of completing your training. It typically involves submitting a well-organized curriculum vitae, an innovative research plan and lining up referees who may be called upon to write reference letters. If you reach the short list, you will be invited to visit the institutions to assess your fit for the advertised position and the institution as a whole. Be very well informed about the department/institutions and show that you are well prepared for the visit. Your scientific talk should be well rehearsed and well presented. An important selection criteria is collegiality and functioning as a team player. Interviewing at a variety of institutions will help you to learn about the interview process and allow you to see what other institutions are offering which will help your later negotiations. If you are invited back for a second visit, be well informed when negotiating salary, benefits, and start-up packages to ensure that you are receiving a reasonable salary package for yourself and a budget that will enable you to establish your research program. Do not rush to hire staff and trainees. Look for the very best and be actively involved in this process. These individuals will be doing much of the work in the lab; therefore, it is imperative that they are of high quality.

*Planning a Career in Biomedical and Life Sciences.*
DOI: http://dx.doi.org/10.1016/B978-0-12-802242-9.00006-2

## 6.1 EXPLORING JOB PROSPECTS

You must explore job prospects well in advance of completing your training. The job search can take more than one year from preparing your application to landing the job and negotiating the start-up package. While many of the steps required to land a job are similar to those discussed in Chapter 5 with regard to obtaining a postgraduate position, for the sake of thoroughness I will repeat some of them here and also note some inherent differences.

Many academic positions are advertised in the fall, with interviews taking place in January through March. As mentioned in Chapter 5, initial contacts can be made in a variety of ways: at scientific meetings, visiting lecturers, job postings in journals and web sites, and through other informal means. Many successful candidates report that word of mouth and networking are crucial to finding out about job prospects. Your supervisor and other department members may also provide you with crucial contacts and information.

Even if you are applying for a position at your current institution or at one you know well, treat the process just as you would if you were applying to an unknown entity. You now have to look at the department or institution from the very different perspective of a prospective faculty member, not from that of a trainee. It will not hurt your cause to interview at several institutions so that you learn about the interview process and find out what packages different institutions are offering. This will help you contextualize the offers you receive and provide you with an indication of what issues are negotiable. Understanding market pressures will allow you to negotiate more shrewdly and more successfully.

## 6.2 ACADEMIC JOB DESCRIPTIONS

Obtaining a faculty position as an independent investigator generally means that you have been assigned laboratory space in your institution to set up a research laboratory. You are now responsible to decide upon, plan, and obtain funding for your research program. You will have to hire technical and scientific staff, and you will begin supervising and mentoring undergraduate, graduate, and postgraduate students. The term "independent" does not mean that you neither collaborate with other scientists nor work within scientific teams and in networks. Team-based research is very important in contemporary biomedical and life science research, and it attracts considerable funding due to its unique ability to tackle important questions, particularly those of a global nature.

## 6.3 ACADEMIC APPLICATION

Remember the search committee will be receiving many applications to review. When you apply for a position, submit a well-organized and unambiguous resume. Make sure to clarify your role in publications, especially multi-authored ones. Do not mix abstracts with publications. List chapters, books, and other nonpeer-reviewed articles separately.

> **Your Research Plan Must Be Well Developed**

Present a well thought out research plan that is not too long, but is innovative in nature and can be feasibly carried out at the institute you are applying to. If you are performing basic discovery research, consider the possible translation and commercialization of your work. Before submitting your research plan, you should obtain feedback from colleagues and mentors about the plan itself and about its long-term prospects.

Do not forget to have a frank discussion with your current supervisor about which projects you may take with you when you leave the laboratory. Be careful about competing directly with your supervisor, as this may lead to great unpleasantness and, in any case, you must establish a research program that is truly your own. While it is wise to keep in contact with, and even to collaborate with, a former supervisor, remember that you need to show research independence.

## 6.4 REFEREES

Pay careful attention to choosing appropriate referees. As a matter of courtesy and for your own sake, let your referees know that they may be contacted. They will be glad to receive your current resume and up-to-date research plan so that they are aware of your current situation and future plans. Referees should know you well, and they should be able to provide critical analysis of your work and of your ability to work in a group. Collegiality is an important factor in choosing both faculty for departments and investigators for industry. If an individual hesitates when you ask for a reference, do not use them as a referee. I once had a student who, year after year, applied for a certain position only to inexplicably be turned down again and again. Through back channels, it was finally learned that one of her reference letters damned her with faint praise. Her fortunes changed with a change in referees approached to write letters of reference. Search committees expect to see highly laudatory letters from those who know you best, including supervisors and former employers. If these are absent or fail to impress, you will need to explain why.

## 6.5 TEACHING

In applying for a job, teaching experience is a plus, but many postgraduate students will not have had much opportunity to teach during their training period. To showcase some of your teaching skills, you will present a seminar so search committees and interested faculty and students can view your organizational and presentation skills firsthand, and witness how you handle a discussion of your work during the question period. You must finish within the prescribed time and leave time for audience questions. Students are members of the search committee so you are evaluated on your teaching and on your comments about teaching and supervision that you make when you meet the student group as well as when meeting with the full search committee.

In order to avoid technical problems during your presentation, make sure to format your talk and check the audiovisual equipment to ensure that you know how to run it smoothly. Ask whether your seminar will be transmitted to other locations and if so, format it appropriately. Many universities require candidates to give a chalk talk in addition to giving a formal seminar, so prepare for this as well, if required.

## 6.6 FIRST INTERVIEW VISIT

You have made the short list of candidates and now you are invited to visit. Usually four to six candidates are invited to visit over a short time span. During your interview, make sure to clearly articulate the nature of the position you are looking for. You may even write this down so as to clarify it in your own mind and ensure that you say exactly what you want to do. Consider beforehand where you can be flexible and where you cannot. Do not under any circumstances modify your own job description to suit your prospective employer's needs during the interview. This tactic raises questions about your own motivation. Conveying the notion that "you will do anything to get a job at this institution" is not a strong selling point at all.

> **Be Well Informed About The Institutions You Interview At**

Learn as much as you can about the department you are planning to visit, and about the parent institution as well, by reviewing web sites, speaking with your contacts and mentors, and reading scientific publications. Being familiar with the ongoing research in the department and the faculties' interests will enable you to discuss issues intelligently with prospective colleagues and identify potential collaborators even before your visit. Also find out what the priority programs are in the department and at the institution, and look into how the institution handles intellectual property and commercialization of discoveries.

Even before you arrive, prepare a set of questions that you need answers to on your first visit. Remember that not every issue needs to be addressed during your first visit and getting into details may be counterproductive. Your main objective is to determine whether this is a suitable place to begin and develop your career and whether this is a city that you wish to live in.

| Fit and Collegiality Are Essential |
| --- |

The search committee will want to determine whether you have what it takes to set up an independent productive research program; whether or not your program fits in well with the department's research and teaching goals and objectives; and whether, on a personal level, you will mesh well with the current departmental faculty. Departments prefer team players with excellent people skills. Demonstrate that you are well prepared for the visit by having done your research and by showing enthusiasm and conveying a sense that you are very interested in the position.

While you should not be shy about discussing your salary, start-up packages, and space, it is best to leave this to the end of your first visit when both you and your potential Chair or Director feel more comfortable with each other. If you return for a second visit, then you will negotiate in much more detail on all matters as noted below. Your requests can be somewhat general at this point, although you should have some notion of what salary you deem appropriate and what you would like included in your start-up package. The conversation is best held in private with your Chair or Director, not with the search committee, and you should make sure to see the space offered.

What you expect to accomplish on your first visit is to ascertain that your fit in the job is suitable for you, and whether you feel confident that you are able to launch a successful career at the institution. The institution will ask the same question, and if the answer is yes, you will likely be invited for a second visit.

## 6.7  SECOND VISIT AND NEGOTIATIONS

After all on the short list are interviewed, the search committee makes a recommendation on who to invite for a second visit. Usually there is an obvious first choice but sometimes a more in-depth look at two or three candidates is required.

On the second visit you are given a thorough visit of facilities, you meet more faculty, especially potential collaborators, and you visit the city, generally with your spouse/partner. Then negotiations can begin in earnest. When you are offered a position, you will be given a deadline to respond. Do not be pressured but remember if you refuse the offer then the institution is delayed in hiring since they need to continue interviewing more candidates. You will be asked if you have other job offers or if you are interviewing elsewhere. Be truthful in your response but you do not need to provide details. These should be considered confidential.

## 6.8  START-UP FUNDS

Your budget for start-up funds will include personnel, equipment, materials and supplies, student stipends, and travel to scientific meetings. You will need to provide the Chair or Director with a list of equipment that you need—either as your own or as communal infrastructure equipment to which you need unrestricted access. Indicate how frequent this access will be and determine what the user cost will be. Your start-up request should not reflect a random specific dollar value, but should depend on what you need and should reflect the true costs at the specific institution. For example, salaries for technical and administrative support staff vary in different regions of the country and in different institutions. Bottom line, you need to make absolutely sure you include what you need to start your program and to work for at least three years without major external funding.

> **Think Carefully About Start-up Packages**

Many faculty members assume they must obtain everything in their start-up package. As I have mentioned, you merely need to obtain what you require to launch your career successfully. Do not expect to be in the same position as a senior faculty member who has been at the institution for a while and has amassed resources. Successful senior faculty have also proven that they are worthy of further support from the department or institution. If you too are successful, there will be opportunities down the line to obtain additional resources and funding. If the Chair or Director gives you the impression that your start-up package is your only opportunity to gain such resources and funding then you may wish to reconsider whether this institution is the right place for you: Does it really offer you the possibility of building your career beyond the initial start-up period?

With this question in mind, investigate external peer-reviewed granting opportunities from relevant agencies. Inquire about internal grant competitions. Does the department or institute provide formal support and mentorship for junior faculty when they apply for initial grant funding? Does the department offer internal peer review and/or grant editing services? Inquire about the availability of internal scholarships or training grants for graduate students and postdocs. Often allocations of these scholarships may be negotiated into your initial contract.

## 6.9 SALARY AND BENEFITS

In order to negotiate your salary successfully, you need to be well informed so that you know what a reasonable salary package is. You can find out what the salary scale is at the institution you are applying to in several ways. In many jurisdictions, the salary scales are listed on public web sites, especially if faculty are officially government employees. Otherwise you may obtain this information from your faculty associations, the human resources department, and/or contacts in the current faculty. Discover what benefits you are also eligible for.

You may indicate that you were looking for a package a bit more generous than the norm and explain why you deserve one. You may also ask if there is any opportunity to do extra work that attracts extra salary. Also clarify how your salary will grow over the years and how this impacts your pension plan. Ask about coverage of moving costs and travel to your new position for you and your family, if applicable. Do not discuss specifics of your offer with future colleagues at your institution. Financial issues are sensitive and are best kept private.

## 6.10 RESEARCH TRANSITION

You should try to ensure that you will have enough infrastructure and supplies on site to start your academic research program as soon as possible after you arrive. Setting up a functional laboratory usually takes longer than you expect. Waiting for labs to be built or renovated and equipment to arrive may delay you considerably. It is also a poor excuse for lack of productivity when applying for research grant funding. To prevent a lull in your publication history, finish as much work as you can in your post-doctoral laboratory so that publications will come out as you set up your own laboratory in the new facility. There may be some overlap as you wind down your postdoctoral position and begin your faculty position, and you may even find yourself commuting for some time. Clear this with your advisor and also ask if you may begin writing grants before you leave your postdoctoral program. Find out when you are needed to fill your new position. Some positions require that you begin on a specific date to fulfill institutional requirements, while others have a somewhat more open-ended start date. These dates may be open to negotiation.

## 6.11 SETTING UP YOUR OWN LABORATORY

Review your start-up budget for the first three years and make revisions as required. Do not forget to include the cost of inflation. Obtain correct personnel salary scales and include benefits. Remember that the renewal of existing staff contracts may result in sudden increases in salary while you are setting up. Find out who are the preferred vendors for your institution and whether discounts are given through the institution or whether you will have to negotiate with the sales people

> **Take Great Care in Setting Up Your Laboratory**

directly. Explore equipment prices carefully and query whether set up and training costs are included. Sometimes demonstration models are available. Meet your neighbors in the department and also in other departments. Find out what they do and what types of equipment they currently have. You may be able to share large pieces of equipment with another lab, thus saving on precious start-up funds, or you may even purchase new equipment together with other faculty members.

## 6.12 HIRING TRAINEES AND STAFF

Your lab will succeed or fail based on the people you hire. Since your students, postdocs, and technicians will be doing much of the work in the lab, it is crucial to attract excellent people who will help you reach your scientific goals. So, do not rush into hiring people. Check resumes and obtain references. Personally interview all your candidates for employment and for training positions. Your institution's Human Resources department will have standard procedures for interviewing potential employees. Always follow them. You should take on students and postdocs who will likely be awarded external and internal training grants. These are very good benchmarks for determining student quality.

## CHAPTER NOTES

_____
_____
_____
_____
_____
_____
_____
_____
_____
_____
_____
_____
_____
_____
_____
_____
_____
_____
_____
_____
_____
_____
_____
_____
_____
_____
_____
_____
_____
_____

# The Next 10 Years in Academia

## Contents

## Summary

Once you get a faculty position it is time to plan for the next step which includes tenure and promotion. Be aware of and utilize department/institution pretenure evaluations to guide you in your academic work. You must be very familiar with the tenure process to understand the timeline and the requirements for a successful tenure application. Engage your chair/director and your mentors in the process to guide your academic choices so that you are well prepared to go forward for tenure and promotion. Make sure you have all your teaching evaluations and identify students who may be called upon to provide referee letters for teaching, supervision, and mentorship. Consider whom you would recommend from the best scholars in your field to act as external referees. Tenure and promotion committees will assess your productivity, funding sources, grants received, the impact of your research, and the success of your teaching. During your pretenure period, be a helpful faculty member, working with the leadership of the department to enhance the research and teaching platforms which you work in. You will be required to take on administrative duties but budget your time carefully since your number one priority is your research productivity and your teaching.

## 7.1 CLIMBING THE LADDER

Once you have gotten your first job, you will doubtlessly hope that you can finally sit on your laurels; however, you must immediately begin planning for the next step. Whether you are in an academic or nonacademic setting, you need to familiarize yourself with the career milestones and the pace you should set to reach them. As you progress through the ranks

*Planning a Career in Biomedical and Life Sciences.*
DOI: http://dx.doi.org/10.1016/B978-0-12-802242-9.00007-4

you will need to continually update and revise your dynamic career plan in response to opportunities that present themselves and new information that you receive.

Famously, the academy is rife with bureaucracy and politics; you will need to understand what role the academic leadership plays in your institution to navigate the shoals in these tricky waters. For starters, you should try to understand your Chair's position because the academic Chair's role differs significantly at different institutions. In some, the Chair is a rotating job which is administrative and more of a caretaker position. In others, the University Chair is appointed after a thorough academic search to translate the university's academic mission of scholarly activity into an operational strategic plan designed to enhance departmental academics through innovative transformative planning and execution. The role of the Chair may also differ in different disciplines. In my own discipline, laboratory medicine and pathobiology, the Chair will also most likely be responsible for directing the operations of a hospital clinical department, for example, the clinical laboratory of a teaching hospital that provides academic laboratory services.

As a junior faculty member, you also need to appreciate the standard fiduciary responsibilities and human resources issues that are the chief concern of the Chair and her administrative team. In the academy, the running of teaching programs and of research infrastructure are of major importance, and the Chair is accountable for the success of these programs. In addition, Chairs are expected to actively participate in fundraising which, you even as a junior faculty, can help with. Furthermore, you will notice that the Chairs will also bring certain intangibles to their position pertaining to leadership, vision, determination, and adherence to the academic mission. Speaking as a former Chair, I can attest that the challenges are numerous and much time and energy is needed to mold a successful department, a process in which you as a junior faculty should actively participate. Any help you can provide the Chair is much appreciated. Warning: Do not become a high-maintenance faculty member; rather, work with the Chair to help her carry out her own responsibilities.

As an aside, during my tenure as Chair I discovered that leadership demands more than just being a manager. Obviously, quality management is essential to running a well-functioning department that interacts well with its partners at the university and other institutions. Management of budgets, educational programs, human resources issues, tenure and promotion processes, faculty, student, and course evaluations are some of the areas that require both excellent administrative and academic management. However, leadership is a different issue altogether.

While many books have been written on this topic, to summarize my own credo, leadership requires a well-focused and clearly articulated vision. A smart academic leader blends her own vision with that of the faculty, so that she will garner maximum support. As junior faculty, participate in these processes since, to some extent, it is your future that is being planned. In reaching the final mission statement or plan of action, compromises will be made; however, certain academic issues cannot brook compromise, lest the academic mission will be jeopardized. What are these two elements? Quality and scholarship.

## 7.2 SCIENTIFIC NETWORKING

Scientists do not work in a vacuum, so get to know and meet the investigators in your field. You will rely on them for advice, collaborations, exchange visits, invitations to speak, and postdoctoral applicants. They will be in a position, if free of conflict, to act as both external advisors and referees. Attend meetings, especially small ones, where it is much easier to meet your colleagues and discuss science. Social settings during a meeting are a very good venue to interact with colleagues. Many a promising collaboration has been set up at these informal social settings. Ensuring that your own trainees are financially able to attend scientific meetings with you is a very good investment of your funds. Your trainees will benefit tremendously and you will be viewed very favorably by grant review and tenure committees.

## 7.3 INSTITUTIONAL PRIORITIES

You will often be told that "there is no space" or "there is no money." This most likely means that your financial requirements or need for space are not high enough on the list of priorities, so the ball is back in your court. You have to make a strong argument backed by quantitative and qualitative data, to have your needs fulfilled. This reminds me of a situation early in my career as a university faculty member. As part of my limited clinical practice, I had a small office which accommodated my needs for my consultation practice in cardiovascular pathology at one of the teaching hospitals. Our hospital department moved to a new location in the same institution and unfortunately there was no office space for me. I was to remain in my original office, which was now outside the confines of my department, at a significant distance from my colleagues' new quarters. This was an unsatisfactory arrangement that defeated the very purpose of my being embedded in the department to provide much needed clinical expertise. I tried several avenues to acquire appropriate space without success. I always received the same response: My need was recognized and legitimate, but there was "no space." Finally, I spoke to my University Chair. He informed the hospital that if "I did not have a place to hang my hat" in the new departmental space, he would reassign me elsewhere. Within a few days, I received space for an office in the department. I learned three lessons: (i) Make a strong argument, (ii) have supportive mentors, and (iii) there is no such thing as NO when it comes to space and resources. It is all about priorities.

## 7.4 ADMINISTRATIVE ACTIVITIES

You will be required to take some administrative duties upon yourself, since everyone is expected to pitch in to help administer the system, but budget your time very carefully. If you are doing your share of administrative work in your department as a junior faculty member, do not hesitate to decline a request to serve on yet another committee but indicate that once your time commitments change you are willing to take on new responsibilities.

**Budget Your Time Carefully**

Remember, in your early years, your number one priority is to build up your laboratory and begin to produce and publish quality research. So in the early years, pick your administrative roles carefully, focusing (both at your institution and externally) on activities close to your research activity, such as graduate committees and scientific meeting program committees. While serving on committees, be an active member and provide constructive advice to support and build your department, institution, or society. This will be appreciated by your colleagues and will offer you the opportunity to participate meaningfully in structuring your own work environment.

## 7.5 ADVOCACY

One of your responsibilities as a scientist is to share your work with the lay public. Recognizing this need, Oxford University has even appointed a Professor for Public Understanding of Science, a position held by Richard

Dawkins from 1995 to 2008. Why is it so important to do this? To make a long story short, this information exchange reminds the public of the essential work scientists are doing to transform the environment we all live in. Sometimes, although the impact is obvious to those in the scientific community, it still needs to be spelled out to the public.

This advocacy initiative is especially important in explaining the importance of basic research in life sciences, for discoveries in this field are often translated into "bedside" applications only years or decades later. For instance, it is important to remind people how Sir Alexander Fleming serendipitously discovered penicillin; how the diagnosis of diabetes or, even, cancer used to be a death sentence; or, how seminal discoveries on blood clotting and thrombosis have prevented and/or reduced the complications of heart attacks and stroke. In many cases, the discoveries in life sciences do not have an immediate application; however, they build up a knowledge base that will quite likely lead to future application. It is crucial to remind people that science offers humanity the opportunity to cure disease, solve global health problems, and lengthen human life, not to mention turning the world into a global village and advancing the spread of knowledge and better living conditions for all.

In transmitting your message, be careful not to oversell your work, but instead be factual and honest in transmitting information to the public. Do not overpromise what science can deliver because this will ultimately result in disappointment, a disappointment that fuels some of the antitechnology and antiscience sentiment in contemporary society. What science does deliver is quite remarkable enough without the oversell! Furthermore, in your advocacy do not fall into the trap of setting up a false dichotomy between science and religion. While some religions or religious denominations may have a difficult time with certain scientific hypotheses or truths, this is not necessarily true of all believers.

Advocating for science is not only a question of supporting your chosen profession. Without being overly melodramatic, it is fair to say that it is a matter of importance for the future of humanity. Furthermore, in some cases it is a patriotic duty, as being at the forefront of science can lead to business and commercial opportunities that no country can afford to pass up. In a fiercely competitive economy, governments and private individuals are forced to decide how to prioritize spending. Populism or misinformation may induce them to cut funding to basic science research and higher learning initiatives. If scientists (who understand science best) do not participate in arguing for the best possible infrastructure, educational initiatives, and research funding to support a high-quality life science enterprise few others will.

How can you become involved in this enterprise? First you need to believe that advocacy is important and decide that you are willing to spend your own precious time on it. Then you need to acquire some skills in advocacy. Most people need to learn these skills because they are often not intuitive. You need to know how to speak to the lay public and to members of government and industry in a way they can understand. The best piece of advice that I have received is to speak simply, clearly, and honestly. I try to present myself as a trusted member of the scientific community who is willing to share up-to-date scientific information for the audience's benefit. I try to tell stories that reflect my own excitement at scientific discoveries that may have direct impact on the community at large and on specific members in my audience. The lay public needs to be convinced that it is in their own best interest to support research and education in life sciences.

## 7.6 TENURE AND ACADEMIC PROMOTION

While each institution has its own tenure/promotion requirements, there are several general principles in the academic environment to consider. First, make sure to understand the tenure and promotion process at your institution. Don't wait to find out what steps you should have taken till the last moment before you go for tenure or a promotion. Second, many academic institutions have mechanisms in place to evaluate your progress while in the tenure stream with a focus on teaching and research (both the quality of your research program and your funding). These evaluations will point out your areas of weakness and provide you with useful strategies to improve your standing. Third, to carefully avoid the many pitfalls and stumbling blocks impeding your progress toward promotion and tenure, consult with your mentors, create a Tenure and Promotion Development Plan, and be fully informed. For instance, be aware of the importance of not only teaching and research but also collaboration, productivity, and awards.

## 7.7 TEACHING

The promotion or tenure committee will take your teaching experience into consideration. In some departments or programs, course teaching may be limited, so you need to gain some teaching experience and an opportunity to be evaluated for teaching through other means. The most highly sought after teaching experience is that of sole course director, preferably of a course you designed, but if this is not possible, even teaching a course others have established is fine. If neither of these opportunities is available, you may prove your teaching ability by giving some guest (invited) lectures in someone else's course. Alternatively, you may gain bonafide teaching experience in your capacity as a graduate supervisor or as a member of a graduate student advisory committee. Mentoring graduate and undergraduate students can also count for teaching if performed in a rigorous fashion with your mentoring strategies informing the process and outcomes being recorded. In my experience as Chair, I had a faculty member hired in a nontenured slot to develop a research program. He made it clear to me that he wanted to teach as well. We finally found him a few guest lectures and some graduate student seminars to supervise. This experience started him on the road to teaching, enabling him to link his successful research career with pedagogy. When he came up for promotion, his teaching dossier had the teaching achievements necessary for promotion, which complemented his excellent research.

Teaching dossiers require a comprehensive description of teaching activities, including course number and title, number of hours taught, number of students, type of teaching, and most importantly, teaching evaluations. These include student evaluations which usually focus on the effectiveness of in-class teaching. Students will focus on the appropriateness of the course workload, the course evaluation processes, the clarity and organization of notes and lectures, the organization of the course as a whole, and your availability to handle questions and problems. They will usually communicate with you electronically or during face-to-face meetings. They will appreciate prompt replies. They will also appreciate the prompt return of class assignments and midterm exams. You should review these student evaluations carefully because they provide valuable feedback that can help you improve your teaching skills. Make sure to collect student teaching and course evaluations after each course you teach. It is very difficult to reconstruct evaluations years after the fact.

Another factor that the promotion or tenure committee will take into account is the quality of the graduate students and postdoctoral fellows you have attracted. Likewise they will be interested in knowing whether you have also been a member of student advisory committees and how your students evaluated you. Graduate students and postdoctoral trainees provide referee letters which often focus on the quality of your supervision and on your performance as a mentor. Your engagement with students as a member of a student advisory committee is important to the tenure or promotions committee, and I have seen very supportive student reference letters commenting on successful faculty performance in that capacity. In addition to student evaluations, some departments/institutions have faculty teaching evaluating committees that perform an in-depth analysis of your teaching, visiting your classrooms, and reviewing teaching material as part of their assessment. Involvement in the various available teaching opportunities will help you improve your teaching and enable you to prepare a strong teaching dossier for tenure and/or promotion.

## 7.8 RESEARCH PROGRAM

First and foremost, you must set aside protected time for your own research enterprise. If you are employed in an academic research intensive institution, you must give your research activities the highest priority. In order to become a successful, independent, well-funded and productive scientist, you must dedicate 75% of your time to research.

> Protected Time for Research

You must choose a research program that is innovative and that will lead to new and important discoveries. "Me too" or "lateral type" research will not advance your career. You need to work on major problems, the solution of which will have impact on your discipline and, if you are lucky, on science in general. You need to choose a feasible project that is doable during your pretenure period. Once you choose your problem, stay focused. Do not hesitate to ask for help or to search for collaborators to complete your work. However, by the same token, be careful not to accept too many offers to collaborate on other projects. This will spread you too thin and you will not have enough time and energy to devote to your own project. Having a wonderful reputation as a collaborative colleague is fine, but you need to show a strong independent focus to compete for tenure.

On occasion, you incidentally discover something very interesting while you are working on your main project. Since you cannot devote time to this as well, you have a dilemma on your hands. Do you put this fascinating and potentially very important finding on the back burner and risk someone else stumbling across it, or do you change projects? If you change projects, you may not complete enough high-quality science before your tenure application must be handed in. Of course, every case will be different; however, it is best to discuss the situation with your mentors and others whose judgment you trust. This is clearly a good news/bad news story, and you need to give it careful and thoughtful consideration.

## 7.9  RESEARCH FUNDING

In terms of research, the tenure and promotion committee will examine whether you have received peer-reviewed funding as a Principal Investigator from at least two funding sources and had your grants renewed. The importance of writing research grants well cannot be overstated. If you cannot obtain funding, you will be unable to perform your research and you will quickly be out of a job.

> **Strive for High Impact Publications**

Start writing early, so that you have enough time to write a thoughtful innovative proposal. Ask colleagues for their input to improve content and readability, and ask them for copies of their successful grants. Take advantage of any assistance your institute provides grant writers with. Be aware that different agencies have different requirements and that they may even have different scoring systems, especially in terms of the weight they give to the grant's components.

Most scientists consider investigator initiated operating grants as the cornerstone of a successful research program, and as essential drivers for new discoveries. Actively seeking collaborations and membership on teams and networks is also important; however, budget your time carefully so that you do not become overextended and fail to do your own research and write your own grants. Being a coapplicant on a multi-investigator program proves that you are making a mark on your field, but you must clearly demonstrate your unequivocal and major contribution to the program. Being a principal investigator on consortium type projects is an excellent career move and does advance your science effectively and efficiently.

## 7.10  ACADEMIC-INDUSTRIAL COLLABORATIONS

Currently, there is much more interest than there used to be in academic scientists collaborating with industry. Although these two cultures still remain distinct, as modern science progresses there are definite benefits in the amount and quality of these collaborations. Such collaboration is immensely valuable because it allows different expertise, technology, and funding to be brought to the scientific table to create new platforms that carry out innovative research. While translational and applied research is often the focus, basic discovery research is not excluded.

Since contemporary governments support innovation in biomedical and life sciences as an economic engine, and as a way to provide high-quality cost-efficient health care, there are substantial funding opportunities for university and industry partnered research. Government agencies in many jurisdictions have developed competitive funding mechanisms to provide research dollars to collaborative efforts between universities and/or research institutes and industry. The two cultures are learning to work successfully together and to cope with the administrative burden that these collaborations often bring with them. Both sides are working on easing these burdens and streamlining the administrative and legal processes for interorganizational collaborative research to occur. Academic institutions are ramping up their administrative capacity to help academics (and themselves) overcome the hurdles and reap the benefits from possible commercialization opportunities that may be derived from research partnerships between academia and industry.

## 7.11 PRODUCTIVITY

Your research program may start off with a few more modest publications to demonstrate initial productivity and independence. However, in order to gain promotion and receive tenure, you must reach a stage where you have had high-impact publications and you have already had more accepted for publication (in press). Quality is far more important than quantity, as you need to establish a reputation among your colleagues for producing a body of innovative high-impact work. Your publications should appear in the highest impact subspecialty and general biomedical and/or life science journals.

These publications should naturally lead to invitations to national and international scientific meetings and to universities and institutes. Invited reviews in well-recognized journals will also indicate your burgeoning reputation in your field. Do not allow flattering letters tempt you to write reviews in less than high-impact journals. These will count little in terms of your scientific productivity and will take you away from important work.

As your reputation grows, you will be invited to review your colleagues' research as an external reviewer of manuscripts and/or grant proposals. Your reputation and standing will be enhanced by the journals that you review for, so prioritize reviewing for high-impact journals and journals in your discipline or areas of research. You may, although this is much less likely, be asked to serve as an internal reviewer on a grant review panel. Many feel that before agreeing to do this, you should wait at least five years, so that your own research program is well established by such a time, and you now have time to devote to the review process. These review activities are time consuming so you must budget your time very carefully. Failing to obtain your own grants because you are too busy reviewing others' will certainly neither advance your career nor help you achieve tenure.

## 7.12 AWARDS

You should certainly apply for and receive personnel awards. Early career awards are usually very competitive and are a confirmation of the high regard your peers have for you and your work. These awards are often meant to identify and support rising stars.

> **Personnel Awards Are Prestigious**

Applying is usually a time-consuming process, so only apply for those awards you have a very good chance of receiving. Don't apply blindly. Find out as much as you can about the application process, so you understand how best to complete the forms and what the agency is looking for in its awardees. Seek advice from senior colleagues and from former award winners. As your academic career evolves and you build a track record of excellence, consider seeking nominations for teaching and/or research awards, both within and beyond the confines of your institution. Such awards not only highlight your own capabilities but place your department, institutions and/or scientific, or professional society in a positive light.

## 7.13 CAREER PROGRESS IN A RESEARCH INSTITUTE

If you have launched your career in a research institute (where there is less of a focus on teaching than in a university) your evaluations will be primarily carried out by your institution, even if you are affiliated or cross-appointed to a university. The standard evaluations are carried out in one of several ways. Perhaps the most common approach is the review of your research after three and six years. Initially you may hold a junior/associate scientist rank and after two successful reviews you may achieve a scientist rank. Some institutions have a further senior scientist rank for their most accomplished investigators. The science is critically evaluated by internal and external experts in your discipline and the same criteria of research excellence that is applied in the university tenure and promotions process discussed above is applied here. Your roles at the institute as a collaborator and as a good administrative citizen also have serious weight. Today, in an era where the value of transdisciplinary science is well recognized, your involvement in successful team efforts at researching the global problems that your institute focuses on will be highly valued.

## 7.14  TENURE AND PROMOTION ACHIEVED

You have finally gained tenure and promotion! What is expected of you now? Once you achieve tenure and promotion, your department and university expect you to continue to dedicate yourself to excellence and to make significant contributions to life sciences and/or biomedicine. However, in order to do this you and your colleagues will need the continued support of your institution and their leadership. You will rely upon them to secure resources to create stable, high-quality environments for faculty and trainees that allow you and your colleagues to continue to learn, produce sterling research, and (where applicable) train high-caliber students, well equipped for careers in both academic and nonacademic life sciences. Such an environment will create a culture of scientific excellence that retains the very best life and biomedical scientists to generate and transmit new knowledge to fellow scientists and trainees and foster innovative and transformative research.

---

### Tenure and Promotion Development Plan
Name:
Department:
Rank:

You should review and understand the tenure criteria before you sign your employment contract, so you know that the tenure expectations can be met by the terms of the contract. Normally, achieving tenure results in promotion as well, so this plan uses tenure to mean tenure and promotion.

This plan should be set up as soon as you begin your academic position. It should be initiated in collaboration with your Chair (or delegate or your direct supervisor). The plan should be referred to frequently and updated and/or amended as required. You should share your plan with your mentors.

Date of Commencement of Position:
Interim Evaluation Dates to Assess Progress:
Submission Deadline for Application for Tenure:

- [ ] I will seek advice from colleagues and academic leadership to plan my tenure application process.
- [ ] I have mentors.
- [ ] I have attended Tenure information seminars organized by the University/Division/Department.
- [ ] I am very familiar with tenure criteria.
- [ ] I am very familiar with the tenure application procedures including deadlines.
- [ ] I have consulted with those faculty leaders who will organize my tenure application, usually the departmental Chair or a delegate.

## Teaching:
- [ ] I am fulfilling teaching assignments (e.g., undergraduate, graduate).
- [ ] I am collecting teaching evaluations, both quantitative and qualitative.
- [ ] I am correcting teaching deficiencies.
- [ ] I can create lists of students and faculty who can assess my teaching.
- [ ] I have had interim teaching assessments from departmental or other committees that carry this out in my institution.
- [ ] I have quality education publications (if applicable).
- [ ] I have course material to provide as part of my tenure dossier.
- [ ] I have a list of students who can provide a reference letter.
- [ ] I have won teaching awards.

## Research:
- [ ] I have an active research laboratory.
- [ ] I have research funding, especially from external peer-reviewed grants.
- [ ] I have high-quality peer-reviewed publications.
- [ ] I carry out collaborative research.
- [ ] I am known widely nationally/internationally for my research work.
- [ ] I have been invited to scientific meetings, universities, and/or research institutes to present my work.
- [ ] I have research awards (young investigator research awards).
- [ ] I have trained and am training doctoral stream graduate students.
- [ ] I have a list of graduate students/postdoctoral trainees who may be asked to provide referee letters.
- [ ] I have a list of well-recognized national/international referees from distinguished academic institutions who are acknowledged experts in my field of study and are free of conflict to provide letters of evaluation of my academic achievements.

# CHAPTER NOTES

# CHAPTER 8

# Into the Future: The Path to Academic Leadership

## Contents

## Summary

Once you have been awarded tenure, you continue your excellent research and teaching. However, you may aspire to holding leadership positions in your department/institution. This also requires planning to first obtain junior leadership positions to provide you with necessary skills to then move into senior positions. As a leader, some of your time is spent for the betterment of your scientific community. You need to maintain your academic standing and lead by example. You should be well aware of your scientific and institutional environment so as to lead your department to the best positions to take advantage of opportunities. You must gain the confidence of your faculty and be an excellent communicator. Mentorship is an important asset in leadership. As you lead, you must be transparent and inclusive, and assess how best to move your department forward and to handle change. The institution has processes that you should follow carefully and resources to help you with a variety of issues including those involving faculty, staff, and students.

## 8.1 ACADEMIC LEADERSHIP

How might you proceed as a senior tenured member of your faculty? You may wish to explore a leadership role in life sciences, perhaps as chair of a department. The current departmental leadership may identify potential leaders who have strong academic qualifications and are willing to subsume some of their own research ambitions to serve the common good. They will be groomed for local succession or for export to leadership positions elsewhere. Generally, they are provided with some suitable administrative experience in the university, initially directing and managing a departmental program. These management roles may include vice chair for education or research, or director of graduate education. Alternatively, the best preparation for becoming a leader in a

*Planning a Career in Biomedical and Life Sciences.*
DOI: http://dx.doi.org/10.1016/B978-0-12-802242-9.00008-6

large department is to have had a similar position at a smaller institution or department. Leaders frequently move from small departments where they "get their feet wet" to larger more complex departments in large institutions.

Traditionally, chairs learned on the job. However, the complexity of the job today and the numerous oversight bodies and regulatory issues have required a change. Academic institutions now usually provide a local curriculum for new administrative faculty that must be learned. In addition, a new chair will benefit from taking advanced courses on university administrative management. New chairs should also take advantage of the more traditional mentorship structure to help them, especially in the first 18 months on the job.

## 8.2 LEADERSHIP'S TWO FACES

Leadership in the academic setting must be directed both outward and inward. The outside community must perceive the chair to be the official leader and spokesperson for the department. The chair must be well known beyond the boundaries of the department and have excellent working relationships with those who are likely to interact with the department.

Since delegating certain duties may send a negative message on many levels to the outside community, the chair needs to be particular in what is delegated. When the department needs to be represented, the chair should speak for the department and present a clear, unambiguous message that reflects where the department stands on certain academic issues. If such a task is delegated, the message may come across as diffuse and confusing.

Within the department, the chair must build positive relationships with faculty, staff, and students. Everybody must know that the chair is there for them, so that when problems arise, they can turn to their leader for a confidential and frank discussion, secure in the knowledge that they will receive useful and readily dispensed advice. A key to creating this rapport with one's subordinates is clear communication that is forthright and honest. Building and maintaining such a form of internal leadership requires sensitivity, and understanding of human nature, and a willingness to spend time to guide and help those in the department. This same internal leadership requires that the chair sets the guidelines, the academic bar, and the accountability framework under which the department operates. These regulations may sometimes appear harsh; however, they should be fair, reasonable, and enacted without creating internal conflicts that damage the fabric of the department/division. Wide consultation and town hall-type meetings that explain the reasons for new regulations and changes will help facilitate smooth implementation.

## 8.3  DELIVERING LEADERSHIP

There is no optimum model; you will develop your own style. You can start by being hands on and learning about how things work in the chair's office. Then begin delegating some authority to your administrative staff and certain colleagues as you see fit. As staff and colleagues prove trustworthy and become familiar with your needs and vision, you will delegate more and more tasks. You should, however, always be in the loop. It should be made clear to both faculty and staff that you must always be informed about current or potential problems. It is very embarrassing and potentially harmful to the department if the first time you hear about a department problem is from an outsider, especially the Dean. Ultimately, there is no substitute for you yourself understanding the complexities of the job and being able to make informed decisions. The "hit-and-miss" form of decision making is aptly named. When the "miss" involves a major issue, you will unfortunately wind up with a bigger problem than the one you originally had.

As a leader, you will gain immensely by (i) expertly using the following paradigm: understand → consult → make a decision → communicate, (ii) assessing risk, and (iii) strategically planning so that your department finds itself in an excellent position as the future unfolds.

## 8.4  CLEAR, OPEN, AND HONEST COMMUNICATION

Keep your one-on-one meetings brief, and make sure everyone is aware of this policy. You will have a tremendous number of responsibilities, and it is crucial that everyone knows that you are happy to give them your undivided attention for a brief, fixed period of time.

Committee meetings should be as brief as possible and never run overtime. If possible, end early; everyone will appreciate that you value their time. Do not, however, inhibit or prevent useful discussion in order to end early or on time; just make sure to schedule sufficient time to hear all opinions.

As department chair, you should treat each and every opinion offered with respect and never dismiss someone's opinion in an offhand manner.

If you disagree and need to rebut on a critical issue, do so immediately in an even-handed way before the meeting is over. Shouting and table banging are not acceptable in the academic community, neither is ridiculing the speaker.

As a chair you must provide and communicate a clear and concise policy that has received approval from the proper channels. The policy must be communicated to all concerned and be easily accessible. Usually policy is fixed and needs approval through governance to be changed. If you successfully make policy changes, remember to write an easily accessible implementation document that comprehensively describes how the policy will be implemented.

## 8.5 AVAILABILITY

For better and for worse you cannot hide today, and you will usually be accessible electronically. However, if you do go away, select a group of senior faculty to whom you can sign out to, who are familiar enough with the day-to-day running of the department to handle nonurgent matters or more importantly, to decide what can wait and what needs your urgent and immediate attention. If you have an administrative manager who has broad knowledge of the department's operation, he will function as a point person or a filter when you are away. When you are absent from the university you must inform both those whom you report to and key leadership colleagues.

## 8.6 SAFETY

The chair must be aware of the physical and occupational health and safety risks that are part of working in a scientific research center. All offices and laboratories must undergo a security review without exception. Universities and hospitals have regulations, guidelines, and resources; however, as chair you have the ultimate responsibility to ensure that these are communicated to your faculty, staff, and students and that they are adhered to.

## 8.7 THE URGENT SITUATION

Leave time in your daily schedule to meet with those who urgently need to see you. By addressing these cases within 24 hours, you can oftentimes avert a more serious problem from developing. Your assistant should be able to distinguish between the truly urgent and less urgent cases. After the initial meeting, you should set up a more comprehensive meeting later at everyone's convenience.

At the first meeting, find out exactly what the problem is and what the individual wants you to do about it. Oftentimes, an individual may just want to keep you in the loop and does not require any action on your part. Assure the person that whatever he or she wishes to keep confidential will be kept confidential. So if you need to consult with others for advice, let the individual know before you do so. If in your assessment safety is or may be an issue, you must act for the common good; however, in any case, let the individual know what you plan to do.

Ask if you may take notes during the course of the meeting, and do so. At the end of the meeting, you should write down the action items that you and your visitor agree to implement. Make sure these are clearly understood, and prepare a timeline for the implementation of all these activities.

Make sure you are well aware of any university and faculty policies concerning the matter under discussion, and, if necessary, consult with others to clarify these. Depending on the circumstance, the institution itself may have very specific guidelines and timelines that must be followed. Direct your interlocutor to these pertinent policies as well, and if the issue is not in your jurisdiction do not make the mistake of handling the case yourself. Guide the individual to the appropriate place to receive help.

If an individual wishes to make a complaint and expects you to take action, you need to receive the complaint in writing and the individual must sign it. This prevents any future misunderstandings and compels the individual to crystallize the complaint. You should make the complainant fully aware of the available options and point out that anonymous complaints are not terribly useful, since they are difficult to pursue. You should also remind the complainant to consult any collective agreements, memorandum of agreements, guidelines for appropriate academic behavior, or any other pertinent institutional policies and documents. While the complaint may make sense to you and the complainant, it may be countermanded by a written agreement or policy under which you operate. So be well informed.

Remember you do not have to handle this alone. There are numerous knowledgeable individuals you can consult. Investigate how the various services at the university may also help. These are usually staffed by professionals with experience in a wide variety of issues, from safety and security to harassment and family and health services. The best approach is to go up the reporting ladder (that is to say, the chain of command). You can maintain confidentiality by seeking advice without naming the individuals involved.

## CHAPTER NOTES

_____

_____

_____

_____

_____

_____

_____

_____

_____

_____

_____

_____

_____

_____

_____

_____

_____

_____

_____

_____

_____

_____

_____

_____

_____

_____

_____

CHAPTER 9

# Your First Job: Choosing and Preparing Well for a Nonacademic Career

## Contents

## Summary

There are many nonacademic career opportunities available in industry, business, and professional disciplines. Some will keep you at the laboratory bench as a research scientist. Others will require that you apply your scientific research knowledge to a nonscientific field such as finance. There are also clinical positions in biomedical health care, such as diagnostic molecular genetics. Depending on your interests, you may become well informed even while you are a graduate or postdoctoral trainee. Before you begin the application process for these jobs, be sure to make appropriate inquiries and make sure you have the required qualifications to work in your field of interest. Do not, however, overlook your academic research; it is an important qualification which provides you with unique skills and a knowledge base. Thus complete a high-quality thesis with high-impact publications. You may find that networking in both academic and nonacademic realms will help you with your job search, as many nonacademic positions are advertised through word of mouth. Internships, placements, and volunteering are good ways of developing these networks for trainees and graduates to gain relevant experience. After you have acquired a position, be aware of the criteria for promotion and how to make yourself more competitive for promotion. Maintain your networks and join scientific and industry societies to keep abreast of the changes and opportunities occurring in your discipline.

Many of you will pursue the nonacademic route since only about 15–20% on average of PhD graduates end up in academia. How do you prepare for a nonacademic job? There is no formula for doing this. While a graduate degree in biomedical and life sciences is an asset for some nonacademic positions and is essential for others, overall your science research will play a large part in shaping any nonacademic career. The career options are

*Planning a Career in Biomedical and Life Sciences.*
DOI: http://dx.doi.org/10.1016/B978-0-12-802242-9.00009-8

extremely heterogeneous, the competition is fierce, and the recruiters are only looking for highly qualified applicants, so you must be well informed about your choices.

## 9.1 TRAINING

To train for a nonacademic career in the biomedical and life sciences, you must remember two things. First, do not let your preparation for this non-academic, complimentary, or new career path interfere with your completing a strong biomedical/life science graduate degree. This degree, at the end of the day, is your most important qualification. Second, make sure to find out what qualifications are required to work in your particular field of interest, such as certification, licensing, interning, and so forth.

Once you have investigated the formal training opportunities available for this nonacademic path, you will have to decide whether to complete this additional training while you are working on your degree or afterward. Some life sciences programs offer combined or dual graduate degrees/programs. Whether you choose to do both concurrently will depend on how hard you want to work, how well you multitask, and whether this is at all feasible given the demands of your degree program and the training required for your nonacademic career path.

We often assume that training entails taking formal courses, which are primarily comprised of lectures and digital modules. And, indeed, the new emerging educational technologies make it easier to gain access to knowledge and engage in independent learning online. However, be careful to ensure that the education you are receiving is of a high caliber.

There are, however, numerous other ways to learn about a new field or discipline. Mentorship from those in the field is to be highly recommended. You can often find mentors through the offices of a scientific or professional society, as these societies are usually interested in recruiting scientists to their fields. They will be happy to offer other resources as well to those interested in exploring their disciplines.

Volunteering a few hours a week has much to commend it. You will experience a new area firsthand. You will make useful contacts, and people in the field will have the opportunity to see your work and be able to provide references down the road. You may also be able to produce some work that could become part of the portfolio that you will use to apply for positions.

## 9.2 LIFE SCIENCE RESEARCH AS A PLATFORM

Your science research training in graduate school may offer you a platform upon which to train in a new field. By combining fields or disciplines, you will attract recruiters who prize such a set of skills, which are likely to foster innovation.

## 9.3 JOB OPPORTUNITIES

If you are interested in health care, you may consider exploring clinical diagnostic fields such as clinical chemistry, clinical microbiology, clinical genetics, and clinical molecular diagnostics. These require additional post-doctoral type training and most often certification by a professional body.

The traditional nonacademic scientist positions in the pharmaceutical and biotechnology industries allow you to either remain a bench scientist or become a laboratory or program manager removed from the bench. In these positions, some of you may collaborate closely with academic institutions through teaching and research. If you carry out scholarly activity, you may even have university appointments and participate in graduate student supervision. The university and industry will need to collaborate to ensure that their institutional requirements do not come into conflict, especially with respect to graduate student training. The interface between academic and nonacademic is becoming less rigid, and there is crossover and crosspollination that often benefits both parties.

In the pharmaceutical industry, there are many opportunities for life sciences graduates including medical science liaison, regulatory affairs, marketing, account management, risk management, medical affairs, clinical research coordinator, and research project management, to name but a few.

Biomedical and life sciences training also provide you with the background to enter numerous other nonacademic fields. If you enjoy writing then science writing, editing, patent writing, or publishing may be an option. If you have a business or financial background or are interested in these areas, you may find the business and financial side of science to your liking. Financial analysts with a background in life sciences play an important role in many financial institutions. If you would like to engage in and affect science policy, you may wish to find employment in government and nongovernmental science and health care agencies and groups.

Teaching science at the high school level is another possibility to consider since this provides you with the exciting opportunity of mentoring teenagers interested in life sciences careers. The very presence of highly qualified scientists in high school life science programs enhances the high school experience, making studying science a more attractive option for inquisitive students.

## 9.4 THE NONACADEMIC JOB HUNT AND PLACEMENTS

The same general principles discussed above concerning your training and your first job hunt in the academic realm apply to the nonacademic one. For instance, completing a postdoctorate or an equivalent program will still give you an edge in competing for jobs, especially the very good ones. Likewise, during your placement and academic training, creating a network for yourself through direct contact with peers, colleagues, faculty, and staff is vital. Word of mouth and creative use of social media are also important tools and often lead to valuable information about job opportunities.

Recruiters and potential employers for non-academic jobs will initially look at your well-prepared resume for an understanding of your academic record that allows the recruiter to evaluate your skill sets and how you problem solve, write reports, and communicate. Possessing people skills and being a strong team player are very valuable assets that you should work hard on developing. Most employers will do reference checks focusing on your work ethic, your compatibility with peers and management, your communication skills, and any strengths and weaknesses you might have that would affect your job performance. Always use social media responsibly since digital information cannot be contained, and tends to live on for a very long time.

Oftentimes, job placements or internships during your academic training may lead directly to nonacademic employment offers. Additionally, while some jobs require you to be job ready, that is, well trained for the position, others provide in-house training. In such a case, your new employer is willing to invest in you: spending resources and time to train you.

## 9.5 HEAD HUNTERS

There are several well-known human resources companies that specialize in providing scientific staff and recruiting services on behalf of various industries, agencies, and institutions requiring life sciences graduates. They service the pharmaceutical and biotechnology industries, government agencies and institutions, contract research organizations, and many others.

It is worthwhile learning about these human resource companies, understanding how they place scientists, and researching what obligations you have to them if you decide to use their services. Read all contracts carefully, and seek advice as appropriate.

## 9.6 INTERNSHIPS

The internship, usually unpaid, or if you are lucky, a paid one supported by agencies, philanthropies, and professional/scientific societies, is fast becoming a common way for graduates to learn on the job and gain valuable experience. You should explore what internships are available in your areas of interest and decide whether this route into the workforce is financially feasible and likely to be helpful for you. You may not land your dream job initially, but work experience is vital. With this as an asset, along with the contacts you make and information you gain during the internship, you may yet land your dream job.

Always keep your entrepreneurial spirit alive! Some graduates use the skills developed in their curricular and noncurricular learning to startup their own businesses or develop a consultancy practice, even during their original degree program. Without work experience, it may be too difficult, though not impossible to take this step so early in your career, but it is an option. If you choose to do so, make sure to inform and discuss your plans with your thesis supervisor.

## 9.7 CAREER ADVANCEMENT

Once you acquire your first position you must learn and understand the criteria for promotion within your company. Be aware of milestones and of what activities and/or additional educational offerings you should avail yourself of to make you competitive for promotion. Understand that in some positions you will actually have to leave your company for an advanced position at another company, since there is no further advancement for you in your current employment situation. Familiarize yourself with the climate in your industry and create and/or join networks and social media links of like-minded folks. Also join professional societies to keep your knowledge cutting-edge and to increase your access to opportunities in your field.

# CHAPTER NOTES

# APPENDIX: COMPREHENSIVE ANSWERS TO FREQUENTLY ASKED QUESTIONS

a. **Are there university life science or biomedical research programs designed for high school students and how might they prove useful to me?**

Yes, there are programs that will provide you with a glimpse of research and university life. If you have an interest, and especially a passion, for the subject area, the experience will be beneficial, especially if you have a sound background in high school biology and science. The experience of building on what you know will be exceptionally exciting.

However, if you do not have any interest in this subject matter, you will likely find the program boring and a waste of time. While being exposed to life sciences in this type of program may foster your interest, especially if you have no other career interests at this stage, it very well may not. If you have other passions, you should follow those. Perhaps the same university offers programs and opportunities in your own areas of interest. I had a very bright high school student attend the laboratory to obtain some hands on experience. However, every lunch hour and even during quiet times in the laboratory he spent his time reading philosophy and history. He loved these subjects and was only in the laboratory because a family member suggested he try out science. While he may have learned about the philosophy and history of science during that summer and gained a well-rounded education, research science was not his particular passion. So if you already have a passion, go for it.

The best programs provide faculty lectures and mentored research opportunities in an active research laboratory. They are usually held during the summer months. To gain the most out of the experience, you should have a well-focused and supervised project which is doable in the brief time you will be in the laboratory. Expect to work hard. This is not a spectator sport. Not all experiments work out, in fact, many do not; however, you will learn valuable techniques and you will learn a valuable lesson: in science (as well as in life) you can learn sometimes from every failure.

During your time in the lab, you should interact with the graduate students—in addition to professor or your supervisor—for they can give you true insight into what training to be a research scientist is like.

You should also learn about a few basic research skills, including laboratory health and safety; keeping laboratory records; how to talk about your scientific work with faculty and peers; and how to write simple reports and present your research in an oral or poster presentation to your peers and faculty. The techniques and conceptual knowledge you will learn will largely depend on the research project you carry out.

For the motivated students, a research program designed for senior high school students at a university can definitely prove inspiring and help define future courses of study and career paths. To find out about possible programs, contact your high school guidance counselors and the undergraduate recruitment sections of the university web site. Note that there are often scholarships available that you can apply for.

b. **Will a scientist at the university help me with my high school science fair project?**

Yes, however you are best served by identifying a professor who has an interest and expertise in your project. Find out through science teachers, web sites, and through word of mouth (siblings, older friends, and so forth) who at the local university is working in your area of interest. The best form of communication is e-mail. Do not send out mass e-mails. Customize your e-mail to a few professors who you think can help. Be polite. Describe the rules of the fair. Provide details of your science so that the professor understands what you want and how much time she will need to devote to your project. Make sure you include the dates between which you will be doing your project. Indicate when the project needs to be finished by, and how you can best be contacted. Suggest a possible first meeting if the professor is interested in helping you.

You can prepare a short abstract that has a title and the following sections: background of the project; materials and methods; explaining how you are going to do the project; results indicating what you are going to analyze once you do the experiment, and conclusions based on the projected results. Here are two sample abstracts from my own publications to give you a sense of how to proceed.

## Abstract
## Wnt3a/β-catenin increases proliferation in heart valve interstitial cells.

Xu S., Gotlieb A.I.

**BACKGROUND:** Valve interstitial cells (VICs), the most prevalent cells in the heart valve, mediate normal valve function and repair in valve injury and disease. The Wnt3a/β-catenin pathway, important for proliferation and endothelial-to-mesenchymal transition in endocardial cushion formation in valve development, is upregulated in adult valves with calcific aortic stenosis. Therefore, we tested the hypothesis that Wnt3a/β-catenin signaling regulates proliferation in adult VICs.

**METHODS:** Porcine VICs were treated with 150 ng/ml of exogenous Wnt3a. To measure proliferation, cells were counted on day 4 posttreatment and stained for bromodeoxyuridine (BrdU) at 24 h posttreatment. β-catenin small interfering RNA (siRNA) was used to knock down β-catenin expression. Apoptosis was measured with terminal deoxynucleotidyl transferase dUTP nick end labeling assay. To assess changes in β-catenin, cells were stained for β-catenin at days 1, 3, 6, and 9 posttreatment. Western blot for β-catenin was performed on whole cell, cytoplasmic, and nuclear extracts at day 4 posttreatment. To measure β-catenin-mediated transcription, TOPFLASH/FOPFLASH reporter assay was performed at 24 h posttreatment.

**RESULTS:** Wnt3a produced a significant increase in cell number at day 4 posttreatment and in the percentage of BrdU-positive nuclei at 24 h posttreatment. The increase in proliferation was abolished by β-catenin siRNA. Apoptosis was minimal in all conditions. Wnt3a produced progressively greater β-catenin staining as treatment length increased from 1 to 9 days. Wnt3a produced a significant increase in β-catenin protein in both whole cell and nuclear lysates after 4 days of treatment. Wnt3a significantly increased TOPFLASH/FOPFLASH reporter activity after 24 h of treatment.

**CONCLUSION:** Wnt3a/β-catenin signaling pathway is an important regulator of proliferation in adult VICs.

From Cardiovasc Pathol. 2013. 22: 156–66

**Abstract**

**Cell density regulates *in vitro* activation of heart valve interstitial cells.**

Xu S., Liu A.C., Kim H., Gotlieb A.I.

**BACKGROUND:** Valve interstitial cells, the most prominent cell type in the heart valve, are activated and express α-smooth muscle actin in valve repair and in diseased valves. We hypothesize that cell density, time in culture, and the establishment of cell–cell contacts may be involved in regulating valve interstitial cell activation *in vitro*.

**METHODS:** To study cell density, valve interstitial cells were plated at passages 3–5, at a density of 17,000 cells/22 mm × 22 mm coverslip, and grown for 1, 2, 4, 7, and 10 days. Valve interstitial cells were stained for α-smooth muscle actin and viewed under confocal microscopy to characterize the intensity of staining. To study time in culture, valve interstitial cells were plated at a 10-fold higher density to achieve similar growth densities over a shorter time period compared with valve interstitial cells plated at low density. α-Smooth muscle actin staining was compared at the same time points between those plated at high and low densities. To confirm valve interstitial cell activation as indicated by α-smooth muscle actin staining, valve interstitial cells were stained for cofilin at days 2, 5, 8, and 14 days postplating. To study the association of transforming growth factor β with valve interstitial cell activation with respect to cell density, valve interstitial cells were stained for α-smooth muscle actin and transforming growth factor β at 2, 4, 6, and 8 days postplating. To study the activation of the transforming growth factor β signaling pathway, valve interstitial cells were stained for pSmad2/3 at days 2, 4, 6, 8, 10, and 12 days postplating. To study cell contacts and activation, subconfluent and confluent cultures of valve interstitial cells were stained for β-catenin, N-cadherin, and α-smooth muscle actin. Also, whole-cell lysates of subconfluent and confluent valve interstitial cell cultures were probed by Western blot analysis for phospho-β-catenin at Ser33/37/Thr41, which is the form of β-catenin targeted for proteosomal degradation.

**RESULTS:** The percentage of valve interstitial cells with high-intensity α-smooth muscle actin staining decreases significantly between days

1 and 4, and at confluency, most cells show absent or low-intensity staining, regardless of time in culture. Similar results are obtained with cofilin staining. Transforming growth factor β and nuclear pSmad2/3 staining in valve interstitial cells decreases concurrently with valve interstitial cell activation as cell density increases. Examining β-catenin and N-cadherin staining, single valve interstitial cells show no cell–cell contact with strong cytoplasmic staining, with some showing nuclear staining of β-catenin, while confluent monolayers show strong staining of fully established cell–cell contacts, weak cytoplasmic staining, and absent nuclear staining. The presence of cell–cell contacts is associated with a decreased α-smooth muscle actin. The level of phospho-β-catenin at Ser33/37/Thr41 is lower in confluent cultures compared with low-density subconfluent valve interstitial cell cultures.

**CONCLUSION:** Cell–cell contacts may inhibit valve interstitial cell activation, while absence of cell–cell contacts may contribute to activation.

From Cardiovasc Pathol. 2012. 21: 65–73

If an active scientist takes the time to offer you help, then you have the responsibility to do the work suggested and to attend any required meetings. If you are going to do any laboratory work in the university laboratory, you must be instructed on laboratory health and safety and provided with any other information necessary for working in the laboratory. You must follow all the laboratory rules. Most likely a student or staff member will guide you in your project.

c. **As an undergraduate, should I seek out several research experiences or should I stick to one laboratory?**
The answer to this question is that it depends on your purpose in carrying out research projects as an undergraduate biomedical or life science student. If, on the one hand, you view undergraduate research as an important prerequisite to securing a graduate position in a high-quality department, then the type of research is not critically important and you may either stay in one laboratory or move around. What is important, however, is performing very well during your project and obtaining strong letters of support from those who now know your research capabilities first hand. If, on the other hand, you view undergraduate

research as an opportunity to begin seriously seeking a research path, then sample different approaches and research questions, attending different projects in a number of laboratories will best suit your needs. To visit a number of laboratories and do several projects, you will need to begin research projects early in year one or two so that you can have time for such a broad experience. This approach will usually not allow you enough time as an undergraduate to do in-depth work.

Finally, if your wish to do undergraduate research stems from a desire to test yourself in the research environment and discover new knowledge that can lead to a substantial published paper, then undertaking several projects for credit and/or summer work all in the same laboratory on the same proposal will provide you with the best opportunity to make this happen. My close colleagues and I have had several students who carried out one project over several mentored research opportunities in the same laboratory, and ended up with a first authored publication. This type of productivity puts you in an excellent position to be awarded a graduate scholarship/studentship. It assures the award committee/awarding agency that you indeed have the ability to study a research question in-depth and produce a first authored paper in a very good to outstanding biological journal.

Ultimately, you must make up your own mind as to how you want to proceed. But, I will note that if I was your supervisor and recognized your passion for life sciences in my undergraduate research program, then I would give you priority to continue in my laboratory. If you chose to do so, you would then be expected to move the project forward, not just performing more of the same experiments but expanding your technical and knowledge bases as you investigated the problem in greater depth.

d. **Should I go directly to graduate school after completing my undergraduate program?**

There is no uniform answer to this question. It depends on your individual circumstances. As long as your decision is well informed then there is no right or wrong choice. There are, however, a number of factors to consider that I will comment on, in discussing the following scenarios.

One common scenario finds you are tired of the academic student grind after completing a challenging undergraduate program. You need some time to collect your thoughts, explore new avenues, and see the world around you from a nonstudent perspective. If you are not motivated at present to continue your schooling, and you do not have an existing passion driving you to continue your formal education, take a break. Use this time productively, while also deciding whether or not to go to graduate school. Mentors and those who have recently gone through the process can provide useful guidance.

Another common scenario finds you in the midst of some exciting educational activities. Your undergraduate program has pointed you to a certain career path, and you are passionate and excited about what comes next. In this case, continuing on is a good idea. You do not have to fear losing momentum and you will be able to derive the maximum advantage from what you have done so far. For instance, if you have already made impressive progress in your research project, you will be ahead of the game as you enter graduate school. This is a strategy that entails going with your strengths to maximize your potential for success.

A third scenario finds you passionate about biomedical or life sciences and motivated to go into that field, but wanting to take some time off to catch your breath. In this case, take some time off, but speak to your supervisors and mentors to arrange your next career move before you leave. This will allow you to maintain career momentum. Try to confirm a position for when you return. Since investigators have to plan ahead to fill their complement of student positions, this approach can be attractive to both you and your potential supervisor.

Another scenario that is more and more common is that of the "mature" graduate student. You may either make a conscious decision to put off graduate school or you may have other obligations or opportunities steering you in another direction for several years. After having been in the workforce and seen the numerous opportunities out there for someone with graduate school education, you now decide that you are ready to return for formal studies. Since you have joined the workforce, you are likely to have become accustomed to a certain lifestyle based on your income. As a graduate student, you are unlikely to match your previous income and maintain your lifestyle. If you plan to work outside of the laboratory to increase your income, plan this carefully and discuss this with your supervisor to make sure you will not be diminishing your research experience and productivity. If you wish to focus on your studies, without other sources of income, you will probably need to adjust your lifestyle. The sacrifices you are making to return to school indicate that you are motivated and passionate about your studies. Your life experience is a valuable asset in working in the laboratory group setting and may aid in the process of moving basic science from bench to bedside. Scientific knowledge is expanding rapidly these days, especially in biomedical and life sciences, so you may have to take some refresher courses before you are eligible to apply for doctoral studies. These will bring your undergraduate education up to date, teaching you new knowledge and techniques.

e. **Is occupational health and safety really important?**
The importance of knowing and implementing occupational health and safety procedures and protocols cannot be overemphasized. All institutions must ensure that their students, faculty, scientific staff and administrators, and secretarial staff are aware of the comprehensive set of regulations in place and have been provided with opportunities to learn and practice them. Students and employees must be trained before they begin work and records of this training must be maintained. This implementation is not voluntary; the administrative process is open to audit by regulatory agencies that have oversight of safety in the workplace. While the institutions usually have the overall liability for workplace safety, in practice the supervisors and managers provide day-to-day oversight and enforcement of the regulations and practices. They must help create and promote a safe work environment culture throughout the institution.

Life sciences and biomedical research laboratories have many potential hazards and each institution provides its employees, students, and staff with information about these hazards and designs mandatory courses to teach occupational health and safety. As noted above, everyone who works in a laboratory must be well aware of these and learn how to remove or minimize the risks. Personal protective equipment should be provided by your supervisor and worn appropriately. For example, laboratory coats, gloves, close toed shoes and protective goggles or splash visors, all keep you safe and should be worn when called for. Knowing when to work in a fume hood; how to use, store, and dispose of hazardous chemicals; and how to dispose of sharp devices is extremely important. If you use biohazards and/or radiation, you will need to successfully complete a course to teach you safe handling and storage of these hazards. You will not only learn how to work safely with these agents but also how to safely dispose of them and how to respond appropriately in case of accidents. Likewise, you must be aware of all fire regulations and strictly adhere to them. There should be no food in the laboratory for obvious health reasons.

**f.  How can e-mail and social media create problems for students?**
You are all very familiar with digital communications. They are fast, they have the capacity to reach large numbers of people, and they may exist forever in cyberspace. You need to understand the consequences of using these valuable communication tools inappropriately. By using these media carelessly, you may offend others or cast yourself in a negative light. More serious transgressions may bar you from obtaining a student position or a job, and you may even be dismissed from your current position of employment and even publicly disgraced. You can even open yourself up to criminal charges. So, think before you send and do not impulsively respond to communications sent to you. Several institutions have guidelines for using digital communication on institutional and company URLs. These must be strictly adhered to. The stakes are high so you must be very careful and thoughtful about what you post on social media, both text and images. If a problem arises, consult your supervisor and/or graduate coordinator immediately.

Why the fuss about what you write in an e-mail or post on social media? While certain remarks might pass muster in a late-night discussion in your dorm room (or not), once written down and distributed to the world at large they may come across as cold, callous, racist or simply unthinking. To a potential supervisor or employer, this sends a clear message: either you are someone they would never want to associate their good name with or you are impulsive and immature, failing to think before you act. Either way these are not very good characteristics for potential employees or graduate and postdoctoral students. In some cases, you may betray a confidence that demonstrates complete lack of understanding of the right to confidentiality and the implications of breaching it (even if it is your own). The contents may indicate that you are a poor judge of what is in poor taste, or they may reveal that you cannot control your anger or rage and are too quick to be provoked and to respond inappropriately.

When using e-mail to communicate in a professional capacity, and I include all categories of students and trainees, keep it short and to the point and do not embellish your communication with jokes or inappropriate comments. These may come to mind as you prepare an e-mail or twitter, but resist the temptation and use the delete button. Remember that e-mails once sent become part of the record, especially for inquiries, legal matters, and similar processes. If you need to communicate highly confidential and private information, flag it as such in your e-mail. However, for such matters, it is best to use the telephone. If it is difficult to schedule a conference call, take the time to set up a time when everyone involved is reachable. Remember digital phones are also not entirely secure; the most effective and secure form of communication remains the face-to-face meeting, especially if you only need to bring a small group together. Given the difficulty of obtaining the optimal setting, set your own minimum standards on how to handle communications, especially for sensitive issues. Do seek guidance from your supervisor, graduate coordinator, chair, program director and/or mentor.

If you use a specific social media provider, carefully read the contract. Pay special attention to confidentiality issues, the integrity of the digital material sent and received, and the retention of content. When in doubt, you should consult with your institutional IT department to obtain expert advice.

g. **Should it be hard to speak frankly to my supervisor?**
As a graduate student, you should be able to speak to your supervisor about anything related to your courses or research. Your personal life is just that, personal. If issues in your life affect your educational activities, then it would be wise for you to discuss the issues, in confidence, with your supervisor before taking any other action. Your supervisor may need to seek advice from others at the university, depending on the issue. This is normally done with your permission. Your supervisor may also refer you to another individual for proper advice, if he deems this necessary.

What may be a difficult problem for you is to discuss a specific issue with your supervisor when it involves your supervisor himself. It is always best to speak directly with your supervisor to try to address the issue. Sometimes what appears to be a major problem in your mind can be resolved quite simply by making some minor changes. Before speaking to your supervisor, you may wish to consult with your graduate coordinator or one of your mentors to seek advice on how to communicate most directly with your supervisor. If at all possible, it is best to make sure that no one approaches your supervisor on your behalf before you have a chance to speak with him, since this may put the supervisor in an awkward and potentially embarrassing position. These issues should be kept confidential so that you and your supervisor can be more comfortable and open in your discussion. If, however, you need someone to mediate an issue or dispute with your supervisor then the most appropriate person to turn to would be the graduate coordinator. The graduate coordinator can and should provide an unbiased and well-reasoned approach to mediating the issue at hand. In terms of the departmental chain of command, if the graduate coordinator is unhelpful or unavailable, in most departments, you will need to meet with the chair, or perhaps the chair's delegate, such as the departmental vice-chair of education.

On the lighter side, should you be able to speak to your supervisor about matters of general interest? Arts and culture? Sports? Movies? Politics? Family life? There is no set rule; however, this largely depends on your supervisor's own personality and interests, how busy they are, and whether they actually like to chat. Some supervisors and, for that matter, peers in the laboratory, like to separate their professional life from their personal home life, and therefore are not open to sharing this information with others at work. This should not affect your workplace interactions and is not a cause for concern. However, this will affect the sense of community in the laboratory and may have some influence on your decision to join or even to stay in a particular lab or with a particular advisor.

h. **What can I do if I realize my supervisor is not the right fit for me?**
The first step to resolving a problem is identifying it, so by identifying a perceived incompatibility with your supervisor that is affecting your graduate work and most likely, your ability to progress, you have taken the first step. Try to further clarify for yourself what the issue or issues are. Is it the supervisor himself? Is it the laboratory group? Is it the department or institution? You should try very hard to determine where the problem lies because this will influence possible courses of action and outcomes. In your own mind, do you feel that changes can be made in your current laboratory and/or supervision that will improve your situation? If the answer to this question is yes, then how do you foresee these changes occurring?

You will probably have to speak to your supervisor in a frank and open manner, express your concerns and offer some suggestions to resolve your concerns. If you cannot come up with a solution or you do not wish to discuss the issues with your supervisor, then you have to talk to your graduate coordinator. She will probably be able to offer possible resolutions you did not consider. She will also tell you what your options are should you decide to leave your current supervisor and laboratory. It would also be wise to consult with members of your student advisory committee before taking any action. Remember that you agreed to work in your current laboratory and your supervisor chose you instead of someone else. You, your supervisor, and your graduate coordinator should try hard to mediate the concerns and solve the problem, as you both have a responsibility to make this relationship work. On the one hand, if you leave, your supervisor will have a gap in the research team, and your departure may cause him to lose time, resources, and momentum in pursuing the research project you were responsible for. On the other hand, your progress through graduate school is of paramount concern, and if your supervisor and you are truly incompatible and no compromise can be reached then your working relationship should be ended. Whatever the result may be, do not allow this process drag on for weeks. Attend to this on a high priority basis and move forward.

Before deciding to call the relationship into question, remember that finding a new supervisor may not be easy due to lack of space or funding. If you change supervisors, you will also probably change projects. Make sure that you understand how much time you will lose and make sure that student stipends are available for you to complete your degree in a satisfactory manner.

That having been said, incompatibility with a supervisor does occur. This is why I have constantly stressed the importance of finding out as much as you can about the laboratory group and the supervisor before accepting a position. Since graduate student positions are very competitive, you may be so grateful to be accepted that you overlook or even worse, disregard warning signs that you are a poor match for your potential supervisor. Unfortunately, this faculty member's supervision style will not only fail to inspire you in your research work, it will be detrimental to your finishing your degree.

i. **Extracurricular activities take time. Are they worth it?**
The transition from high school to college or university is not easy. In particular, a curricular load that allowed for and encouraged extracurricular activities in high school is replaced by one that is heavier and requires careful time management. Even so, most college or university students manage the transition and find that the extracurricular activities they have time to participate in provide for enjoyable social interactions, in addition to a healthy lifestyle. Student clubs may even be educational as they foster new interests and provide opportunities for exercising people skills, learning team building and networking.

The life of a graduate student can be lonely. You work on your own project and you have to deal with your own individual successes and failures. Progress may be slow, and you may need to expend much effort before you see any positive results and feel that your hard work is paying off. Extracurricular activities provide several advantages: a change in focus from the research to yourself and to your community; positive feedback over shorter time spans that help to compensate for the longer time frames required to achieve success at the laboratory bench; and individual and team sports that allow you to get some exercise and meet people outside your own lab group. The university also offers numerous clubs and associations students may participate in, as well as the opportunity to be active in student and/or university governance. Departments even have undergraduate and graduate student organizations that provide social and academic programs for the benefit of their fellow students.

Those involved at the leadership level of student governance gain much from the experience. They learn how to work together on a team and how to administer programs and events. I have seen successful student groups put on research days: organizing the poster boards, the faculty judges, the refreshments, and the invited speakers. In many ways, designing research days is the perfect extracurricular activity, as this trains the students for their future in academia, where they will be asked to organize scientific meetings. Some of our student groups have organized scientific seminars with invited national and international speakers. They have created meeting booklets containing the program, the speaker resumes, and much more. These activities are not for "credit" but the experience is invaluable and the sense of satisfaction in doing a good job is very gratifying.

Many students also find an outlet in outreach programs, often designed to help less advantaged youth groups. These programs awaken in these youths the notion that attending university is possible for them. The undergraduate or graduate students may provide mentorship, and even remediation, especially in science, biology, and mathematics.

The trick is balance. Extracurricular activities are to be encouraged and provide many benefits on both personal and educational levels. However, they should not take over a student's life, impacting negatively on academic progress by taking too much of the student's time and distracting him from his course work and research. One of my students has reported that maintaining this balance is a constant struggle.

One of my own peers was a superb and committed student leader. He became so heavily involved that he ended up failing a year in his professional program. This is not to be emulated, so balance your time commitments carefully.

How do you know when you have become overcommitted to extracurricular activities? Students tell me that they know when they suddenly become too busy and feel the pressure of not being able to fulfill either their academic or their extracurricular responsibilities. Feeling rushed all the time takes the fun out of extracurricular activities, turning an asset into a liability. One of my students handles this when she feels it coming, by stepping back and reevaluating her extracurricular commitments.

Institutions do recognize extracurricular activities that are of benefit to fellow students and to the community. There are annual departmental and institutional awards that acknowledge achievement in these areas. These awards are an important part of a stellar resume that impresses academics and CEOs when hiring faculty and employees.

j.  **What is my responsibility as a student to maintain research integrity?**

Research integrity refers to conducting, reporting, and publishing research work that is honest and has not been tampered with thorough fraud, falsification, and/or plagiarism. Honesty in your research program and in your colleague's programs is an important cornerstone of the biomedical and life sciences research enterprise. Each university and research institute has published policies on research misconduct and you must be familiar with them. Loss of integrity undermines the public's confidence in science and leads to severe consequences for the perpetrator. Penalties for broaching integrity may be imposed by the university or institute, a granting agency, or if criminal behavior is involved, a court of law.

Therefore, your laboratory records, including all of your raw data, must be available in an organized fashion so that they can be easily and quickly reviewed. Explanatory notes that you write during experimentation must be included since they clarify issues in your protocol, experimental work, and in your data analysis. Not only are you responsible to maintain the integrity of your research work, you are also responsible for ensuring that colleagues, peers, students, faculty, and laboratory staff also adhere to the codes of behavior for research integrity. If you suspect the code has been breached, then it is best to first seek advice from your thesis supervisor or (if this is impossible because you find the supervisor's behavior suspicious) your graduate coordinator. Make sure of your facts before registering a complaint. Making frivolous or malicious complaints is a serious infraction of codes of behavior and must be avoided. In dealing with any allegations of research misconduct, the authorities tasked with addressing such allegations at the university or research institute must follow the policies in place that are well known to the research community. All communication should be handled with strict confidentiality and privacy. Research integrity is taken seriously by all universities and institutions, and thus it is everyone's duty to maintain research integrity and be vigilant of their own research and that of others.

**k. Can I move from industry back into academia?**

My automatic response is that no pathway you choose is fixed in stone. You are not burning your bridges whether you chose academia or nonacademic career paths. In the biomedical and life science areas and disciplines, when you move from industry to academia and vice versa, there are advantages and disadvantages. On the plus side, you are offering a different way of approaching science and a new and different knowledge base and technical skill set to your potential employer or supervisor. This is an excellent recipe for innovation and will be recognized as such by those who hire you, and should be obvious to you as well. On the negative side, you may feel uncomfortable in your new field at first since you lack certain knowledge. Do not worry about this because as a self-learner and someone looking for new challenges, you will quickly fill in these gaps.

In traversing across the academic–nonacademic divide, you need to be ready and willing to merge the two cultures. Do not try to impose one culture on the other as this will be frustrating and nonproductive. The cultures are different, and moving from one to the other should give you the opportunity for healthy change, not the opportunity to renovate your new position to look like the one you just left. What you should be seeking is a means to blend the two so that you have an improved platform to initiate your work.

The paradigm of turning research scientists into business entrepreneurs is gaining much traction in our day and age. This process began spontaneously as life scientists made discoveries in their laboratories that had obvious commercial potential in the health care sector. For the early pioneers, it was hard to make this transition from scientist to entrepreneur since none of their schooling had prepared them (or their potential partners in industry) to develop and commercialize such biological products. The situation has now changed as university and research institutes offer students and young entrepreneurs, courses, programs and mentoring opportunities to develop an understanding and facility with business entrepreneurship. Thus for many, the academic–nonacademic boundaries are blurring and work at this interface is doable and encouraged in many jurisdictions. Thus scientists are familiarizing themselves with the concepts and terminology of intellectual property, business plan development, value propositions, networking, and general management principles.

# EPILOGUE

My wish is that this book will assist you, the student, with the help of your supervisor, mentor(s), and/or high school guidance teacher to successfully join the exciting community of researchers and professionals working in the academic, industrial, business, and/or professional realms of biomedicine and life sciences. The book is meant to provide you with the tools to effectively plan ahead to achieve a satisfying and successful career.

Your love and passion for biomedical and life sciences is only the first step on a long road. To find your calling among the broad array of careers that biological and medical sciences offer will require careful research and planning; well-conceived career development plans at each stage of your training will help turn your interests and passions into concrete options. Remember, however, to keep these plans flexible so that you can take advantage of serendipitous opportunities that arise. With proper mentorship and timely career planning, I trust you will find your niche in the international community of like-minded scientists for whom science is fun, and for whom science fulfills many of their intellectual, social, and humanitarian needs.

As I conclude this volume, I cannot help but mention that nothing lights up an educator's day more than a thank you or acknowledgment from former students and mentees. Do not be hesitant about letting your teachers and mentors, among whom I include myself, know how you are doing as you move through your training and into the workforce.

Welcome to the community of life science and biomedical scholars and practitioners and all our best wishes for a successful career.

**Avrum I. Gotlieb**

# INDEX

Printed in the United States
By Bookmasters